Oil Palm Fertilization Guide

Bernard Dubos, Xavier Bonneau and Albert Flori

Éditions Quæ

Savoir-faire Collection

Books published in English by éditions Quæ

Evolving the Common Agricultural Policy for Tomorrow's Challenges
Cécile Détang-Dessendre, Hervé Guyomard (eds) 2022, eBook.

Agroecology: research for the transition of agri-food systems and territories
Thierry Caquet, Chantal Gascuel, Michèle Tixier-Boichard (eds) 2020, 96 p.

Artificialized land and land take. Drivers, impacts and potential responses
Maylis Desrousseaux, Béatrice Béchet, Yves Le Bissonnais,
Anne Ruas, Bertrand Schmitt (eds) 2020, eBook.

Can organic agriculture cope without copper for disease control?
Synthesis of the Collective Scientific Assessment Report
Didier Andrivon, Isabelle Savini (eds) 2019, eBook.

Innovation and development in agricultural and food systems
Guy Faure, Yuna Chiffoleau, Frédéric Goulet, Ludovic Temple,
Jean-Marc Touzard (eds) 2018, eBook.

To cite the book:
Dubos B., Bonneau X., Flori A., 2022. *Oil Palm Fertilization Guide.*
Versailles, éditions Quæ, 82 p.
The book is an English translation of:
Dubos B., Bonneau X., Flori A., 2020. *Piloter la fertilisation du palmier à huile.*
Versailles, éditions Quæ, 88 p.

This publication has been financed by Cirad and PalmElit SAS.

Éditions Quæ
RD 10, 78026 Versailles Cedex
© Éditions Quæ, 2022

ISBN paper: 978-2-7592-3675-6
e-ISBN (pdf): 978-2-7592-3676-3
x-ISBN (ePub): 978-2-7592-3677-0

www.quae.com – www.quae-open.com

Contents

Acknowledgements

We should like to thank Cécile Fovet-Rabot, without whom this guide would not have existed, for her critical revision of the initial versions and her suggestions that helped to improve the layout of the guide and its readability.

A big thank-you also to Bruno Rapidel and Alain Rival for their unreserved support and the time they devoted to their careful reading of the manuscript.

We also thank Peter Biggins, longtime translator for tropical oil crops researches, for the translation of the French version into English.

Preface

It is a real pleasure to write a few words to introduce this book. Even if you are not an agronomist, you are bound to feel more knowledgeable after reading this guide. Its originality lies in its not being a long list of recipes. I think the authors' intention is to make everyone aware that oil palm nutrition is a science, and to set up a fruitful dialogue between plantation agronomists and managers, and scientists.

As a breeder, the interaction between plant material and nutrition is one of my constant concerns. Variation in the mineral signature of the leaflets of different varieties is a fact, and agronomists will have to pay more attention to changes in mineral contents (leaflet, rachis), rather than just to a "critical level". The authors do not forget that managing soil fertility also means managing the determinants of soil structure, such as organic matter content, or the soil exchange capacity, which are factors that go hand-in-hand with mineral nutrition. Of course, the reader will gain a better understanding of positive or negative interactions, or competition for uptake between minerals. However, things remain complex and oil palm nutrition specialists will continue to be of great help.

The current commodity price crisis, which follows on from the one in 2011/12, and will be followed by others, challenges us on nutritional efficiency. One of the first keys is how plant nutrition is managed. Numerous publications have shown that nutrition methods based on the "reimbursement" of stocks, exports and leaching of minerals generally lead to an overestimation of real needs. The method presented here is based on long-term experiments and helps to determine actual needs. Such an experimental network should accompany any oil palm nutrition policy. It should not be seen as a constraint but as an opportunity to manage oil palm nutrition in a sustainable and efficient way based on scientific facts.

Another challenge is to define an economic optimum. Fertiliser prices vary, storms are followed by calmer periods, but the trend is towards higher nutrition costs because world stocks are sometimes limited, or prices are strongly linked to energy. Moreover, oil palm responds to fertilisers over the long term: today's nutrition will have an impact on yields in the years to come. There is true know-how to be developed to mitigate costs and adopt long-term nutrition policies that are in line with long-term economic trends.

The approach described here might seem to be reserved exclusively for large plantation companies. In fact, it seems quite possible to make general recommendations for smallholders based on fairly large agronomic units (soils, general environmental conditions) that can be implemented by State Agricultural Development Services.

I should like to sincerely thank the authors for their efforts to sum up decades of experiments and analyses in a short, easy-to-read book that provides an understanding of the underlying evidence-based decision-making approach to managing oil palm nutrition.

Tristan Durand-Gasselin,
senior oil palm breeder and CEO of PalmElit,
a Cirad subsidiary specializing in the development
and diffusion of sustainable oil palm varieties.

Introduction: context and purpose of this guide

With yields 4 to 10-fold higher than other oil crops, and its competitive production costs, the oil palm has become the world's leading source of vegetable fats and oils. These advantages explain the steady increase in areas planted to oil palm to meet growing world demand, especially in emerging and developing countries (Southeast Asia, China, India, Africa).

However, an increase in oil palm areas leads to environmental and social conflicts each time it destroys tropical forests and biodiversity.

Better yields through genetic improvement and appropriate farming practices help to satisfy demand while limiting deforestation risks. Fertilizing oil palm plantations (either inorganic or organic fertilizers) has long been considered a major way of increasing productivity: indeed, it was considered that nutrients should never be a limiting factor and high fertilizer application rates were sometimes recommended (box 1 and figure 1). Today, best environment-friendly practices, and the attention paid to agricultural input costs, call for rational fertilization based on a precise diagnosis of requirements in each cultivated plot.

A full oil palm life cycle involves a continuous increase in frond length up to 12 years old, and in stem growth, with the ultimate height determining the end of the palm's working life and the programming of a new cycle. These specifics have to be taken into account for fertilization decisions over the different periods of the life cycle (figure 2).

Backed up by sound technical information from many years (40) of multi-site trials, this guide is designed to help agronomists in charge of designing fertilization programmes. For each plantation, it proposes fertilizer recommendations that take into account the specificities of oil palm plantings, based on an interpretation of leaf analysis (LA) results (box 2).

Contents depend on the fertilizers applied, but also on other factors (climate, soils, planting material).

This guide proposes to fine-tune the leaf analysis tool by improving each stage of its operation:
– by standardizing whatever can be, i.e., the sampling procedure, sampling period, choice of laboratory,
– by making pragmatic and non-systematic choices for structuring plantations in "leaf sampling units", and the positions of the palms to be sampled,
– by interpreting leaf analyses according to optimum contents determined by fertilization trials conducted in the same soil, climate and planting material context.

Box 1. Fertilization and productivity factors

Fertilization is just one of the productivity factors that determine oil palm yields (figure 1). For a given planting material, bunch (FFB) production is mostly determined by photosynthesis efficiency, which can be limited by the water balance (soil dryness reducing gas exchanges) and insolation (insufficient sunlight, particularly when water supplies are satisfactory). Foliage status is also an important factor, with severe defoliation by insects affecting the production of photosynthates (sugars resulting from photosynthesis).

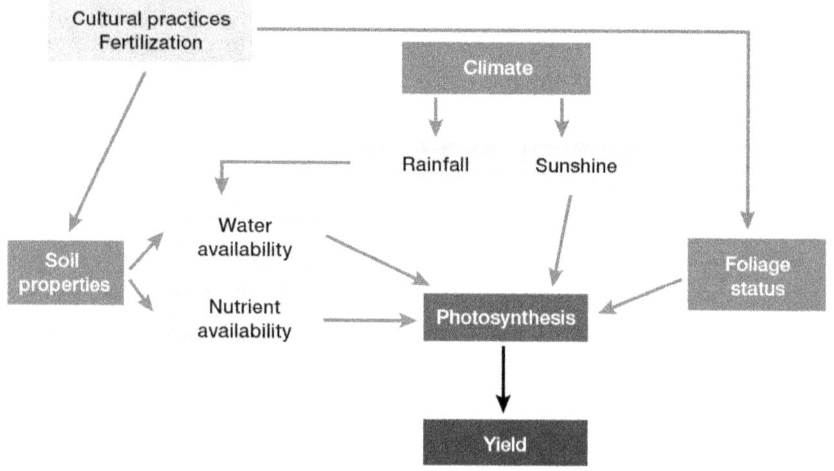

Figure 1. Simplified diagram of oil palm bunch yield build-up

Agricultural practices, including fertilization, along with soil properties, climate data and foliage status affect photosynthesis, hence bunch production. This diagram does not take into account the response time of around two years separating stress periods and their effects on yields.

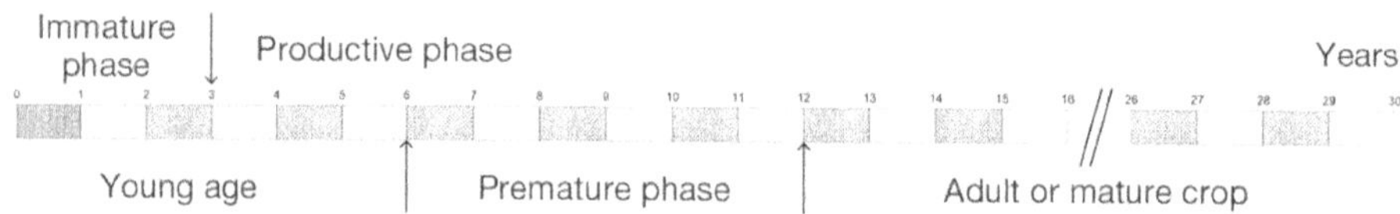

Figure 2. Standard diagram of an oil palm working life cycle

After planting in the field, the first 30 to 36 months are the immature phase, where small bunches are not harvested for economic reasons. Young palms are fertilized according to a schedule specific to each plantation. The productive phase begins after three years with the start of harvesting, which coincides with the start of stem growth. Leaf samples are taken during the productive phase to guide fertilization in the plots. A good indicator is frond length and biomass. At the young age up to six years, growth is highly sustained until the tips of horizontal fronds reach those of neighbouring palms. Frond length increases moderately between six and 12 years, in the so-called premature phase. From 12 years onwards, it is considered that frond biomass is stable: this is the adult (or mature) period, which lasts up to the point where it is no longer possible to harvest bunches once the stem reaches 12 metres in height, usually between 27 and 30 years old. Bunch yields usually reach maximum in the premature phase. They remain stable during the adult period, sometimes with a slight decrease after 20 years.

These well thought-out and standardized protocols guarantee quality analysis results to manage areas that are as uniform as possible, and they provide a relevant interpretation of leaf contents for making fertilizer recommendations.

The guide explains how to compile fertilizer schedules based on fertilization trial results and why they need to be validated by monitoring responses in plantation leaf sampling units.

It also explains why fertilizing oil palm plantations must not be construed as a simple need to adjust a factor that might be limiting. In fact, trials help to define application rates that maintain or raise leaf contents, and they help to fix thresholds beyond which it becomes pointless applying fertilizers. Providing tools for precise and environment-friendly fertilization contributes to society's sustainability expectations. It involves seeking an economic and environmental optimization of fertilization practices.

Box 2. Leaf analysis: a much-used tool

Leaf analyses have been widely used since 1950 to fertilize large agro-industrial estates. Mineral nutrient contents are expressed as a function of the dry matter (dm) weight of leaflets (as a % for N, P, K, Ca, Mg, Cl, S, in parts per million (ppm) for trace elements). Analysis results are compared to reference values, so that fertilization can be adjusted each year according to the division of the plantation into leaf sampling units (LSU). Leaf analysis is a much-used decision-support tool, but it is considered an empirical method. Nonetheless, it remains very widely employed due to its simple use and the quality of the information it provides.

Table 1. Content ranges commonly measured in the leaflets of rank 17 fronds

Nutrient	Symbol	Contents
Nitrogen	N	2.40 - 3.00% dm
Phosphorus	P	0.15 - 0.17% dm
Potassium	K	0.70 - 1.00% dm
Calcium	Ca	0.25 - 0.70% dm
Magnesium	Ca	0.18 - 0.22% dm
Chlorine	Cl	0.40 - 0.70% dm
Sulphur	S	0.18 - 0.23% dm
Boron	B	8 - 15 ppm
Copper	Cu	5 - 15 ppm
Zinc	Zn	15 - 40 ppm
Manganese	Mn	100 - 600 ppm

Understanding oil palm mineral nutrition and diagnosing nutritional needs

The mineral nutrients needed to achieve high yields are taken up differently within the palm's environment and their contents differ in the various organs of the plant.

When these essential nutrients are not available in sufficient amounts, deficiencies occur. Deficiency symptoms usually occur on small groups of oil palms and their intensity varies from one palm to another. However, field observations are not enough to anticipate and apply corrective fertilization on the scale of a plot or leaf sampling unit.

Leaf analyses are widely used to check and guarantee adequate nutritional status. However, leaf contents can be affected by many factors independently of fertilization and those factors have to be taken into account when adopting standards for interpreting contents.

Why fertilize oil palm plantations?

Mineral nutrition is satisfactory when the nutrients needed for the proper physiological functioning of the palm are taken up in sufficient amounts to achieve potential yields at each site. If not, a deficiency occurs; it may or may not have visible symptoms and it gradually limits yields. The intensity of a mineral deficiency depends on mineral reserves in the soil, which are linked to its texture and to the properties governing its storage capacity (cation exchange capacity (CEC), organic matter content (OM), types of clay), and also on fertilizer applications carried out in previous years.

Fertilization trials have shown that a deficiency occurs gradually and gives rise to a significant drop in yields after a few years. This response time varies from site to site, as it depends both on soil reserves and on exports by the crop through biomass production. Oil palm plantations therefore need to be fertilized before the crops become deficient, the aim being to maintain an optimum crop composition enabling expected yields to be achieved at an acceptable economic and environmental cost.

Fertilization effects also depend on other environmental variables, which can become limiting. In highly suitable situations (adequate sunshine, no water stress, low parasite pressure), as in certain regions of Southeast Asia and central America, correcting deficiencies by fertilizing leads to high yield increases, as the other factors are not limiting. On the other hand, in less suitable situations the fertilizer effect is often masked by the impact of other factors. For instance, in Benin, with an average annual rainfall of 1,300 mm (a limiting factor compared to the regions mentioned above), bunch yields are limited to 12 tonnes/ha/year by water deficits of between 600 and 800 mm/year.

In fertilization trials, interannual variations in yields are often found to be greater than the differences observed between treatments with or without fertilizers (figure 3). These variations from one year to the next also occur in well fertilized plots. They are attributed to uncontrolled productivity factors, notably climatic conditions.

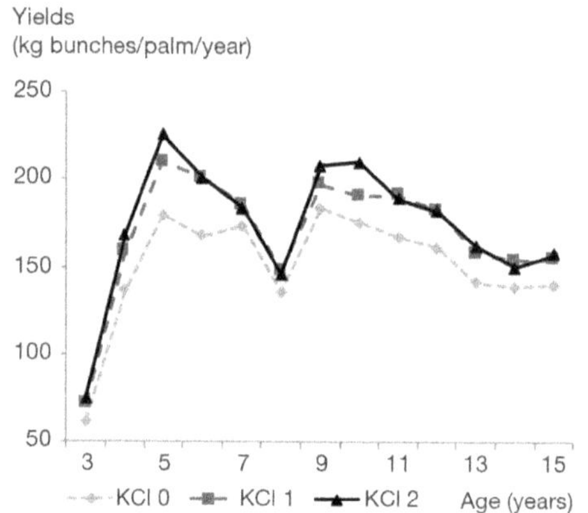

KCl 0, no fertilizer; KCl 1, 1.5 kg/palm/year; KCl 2, 3 kg/palm/year

Figure 3. Example of interannual variation in yields depending on three potassium chloride application treatments in Peru

In this fertilization trial in Peru studying three potassium chloride application treatments, a drop in yields was recorded for all treatments at 8 years. The results showed that the KCl 1 and KCl 2 treatments led to an increase in yields (significant improvement between 5 and 7 years after planting, then again from 9 to 15 years). However, that effect disappeared when another productivity factor became limiting: a bunch pollination defect attributed to a decline in pollinating insects.

Whether for the species *Elaeis guineensis* or for the hybrid *Elaeis oleifera* × *Elaeis guineensis* (called "O × G"), the major nutrients that need to be regularly provided to maintain yields as high as local conditions allow can be classed in the following decreasing order of importance:

Potassium (K) > Nitrogen (N) > Magnesium (Mg) > Phosphorus (P)

This order of importance is found in the quantities present in oil palm vegetative organs (table 2) and fruit bunches (table 3). Chlorine (Cl), a nutrient that is often overlooked, occupies an intermediate position between potassium (K) and nitrogen (N) in the plant's tissues, but it is not systematically analysed.

From 12 years after planting, the biomass of the leaf crown stabilizes and nutrients contained in the fronds are returned to the soil during pruning. Mineral nutrient needs therefore then correspond to an increase in stem biomass and root biomass, but the latter is difficult to assess.

Table 2. Dry biomass and mineral nutrient quantities (kg/ha) in the aboveground vegetative organs of the oil palm (Planting material: *E. guineensis*, origin CIRAD; O×G hybrid, origin Coari×La Mé)

Material	Vegetative orga	Dry matter (kg/ha)	N (kg/ha)	P (kg/ha)	K (kg/ha)	Mg (kg/ha)	Cl (kg/ha)
Côte d'Ivoire 15-year-old *E. guineensis* plantation	Stem	28,228	208	20	386	40	243
	Crown[1]	19,022	186	19	198	34	191
	Total	47,250	394	39	584	74	434
Ecuador 14-year-old *E. guineensis* plantation	Stem	31,309	221	14	407	28	274
	Crown	14,845	225	17	238	21	218
	Total	46,154	446	31	645	49	492
Ecuador 13-year-old O×G hybrid plantation	Stem	12,736	178	22	106	16	36
	Crown	23,296	250	26	287	52	253
	Total	36,032	428	48	393	68	289

[1] Based on 35 functional fronds per crown

Potassium is the main nutrient exported by bunches (table 3). Bunch analysis is a laborious process, which explains why few results are available. Depending on where bunches come from nutrient contents can vary considerably, so estimating the quantities of nutrients that need to be replaced based on a standard bunch composition is somewhat haphazard. The data provided by Ng and Thamboo are those most often used, even though their analyses are old and they characterize a dura type planting material that is no longer topical.

At the mature age, the fertilizer rates commonly applied to compensate for exports vary depending on plantations and continents:
– N, 0.5 to 1 kg/palm/year
– P, 0.11 to 0.22 kg/palm/year
– K, 0.5 to 1 kg/palm/year
– Mg, 0.1 to 0.2 kg/palm/year.

In Malaysia and Indonesia, where yields can reach and exceed 30 tonnes FFB/ha/year, it is common to use 8 to 12 kg of fertilizer (total N, P, K and Mg)/palm/year. Applications are lower in Africa and South America (3 to 7 kg/palm/year).

These differences raise questions as to the appropriateness of standard application rates at each site. Are the maximum rates not too high? Might they be harmful for the environment? What information should be trusted to organize fertilization in space and time to prevent deficiencies from occurring and limiting yields?

Table 3. Quantities of mineral nutrients (kg/tonne FFB) exported by bunches, according to different authors

Reference	N (kg/t)	P (kg/t)	K (kg/t)	Mg (kg/t)	Planting material
Tarmizi A. M. et Mohd Tayeb D., 2006 Malaysia	3.10	0.37	3.92	0.68	*E. guineensis tenera*
Ng S. K. et Thamboo S., 1967 Malaysia	2.94	0.44	3.71	0.82	*E. guineensis dura*
Teoh K.C. et Chew P.S., 1987 Malaysia	-	-	4.32 – 5.12	-	*E. guineensis tenera* on two soil type
IRHO La Mé, non publié Côte d'Ivoire	5.70	0.81	4.26	1.19	*E. guineensis tenera*
Rincón Numpaque A. H. et Torres Aguas J. S., 2015 Colombia	2.92	0.44	3.53	0.58	O×G Coari×La Mé origin

Can deficiency symptoms be trusted to recommend fertilizer applications?

The most frequent mineral deficiency symptoms in plantations are those attributed to N, K, Mg and B. Most have been described in several books and articles: Fairhurst and Caliman, 2001; IRHO Advice Notes (CIRAD-IRHO, 1969; 1991; 1992).

For some nutrients, deficiency symptoms can be unequivocally identified, but that is rarely enough for making precise fertilization recommendations on a plot scale.

Nitrogen (N) deficiency

Nitrogen (N) deficiency causes a diffuse discoloration of the foliage, which turns yellowish green (photo 1). Its detection in mature palms is often subjective and depends on lighting conditions, which can limit the analysis. A deficiency is especially detectable in young plantations, up to six years old, when it is exacerbated by factors that limit nitrogen resources in the soil: excess water, a lack of legume cover crops and high grass density. These problems therefore need to be dealt with at the same time as fertilization is strengthened.

Potassium (K) deficiency

Potassium deficiency is usually described by two types of symptoms:
– appearance of yellow stripes along leaflet margins that gradually spread to the entire lamina, known as mid-crown yellowing (photo 2),
– appearance of small orange spots contrasting with the green colour of the leaflets, known as confluent orange spotting.

These two types of symptoms are found in fertilization trials after several years of potassium (K) depravation, but their interpretation in commercial plantations is

Photo 1. Nitrogen deficiency recognizable by the pale green foliage of two-year-old palms in Côte d'Ivoire

The presence of ferns and waterlogging of the soil in the rainy season reduce the nitrogen resources available for the crop and slow down palm growth.

Photo 2. Mid-crown yellowing on smallholdings in Côte d'Ivoire

A lack of fertilizer applications over many years leads to potassium deficiency which, with these symptoms, reduces bunch yields.

tricky. The yellow stripes only occur when leaf contents have reached extremely low levels (< 0.50%); such a situation is rare when plantations are fertilized, even if applications are not enough to achieve the best yields. Confluent orange spotting can be confused with other damage: foliage fungi such as *Cercospora* sp., insect or mite bites, abnormalities of genetic origin. Visually diagnosing a potassium (K) deficiency is tricky because it can be concluded that nutrition is satisfactory when it is not and, conversely, that there is a deficiency when there is none.

Magnesium (Mg) deficiency

Magnesium (Mg) deficiency is usually more spectacular than harmful. The yellowish-orange zones most often located at the tips of fronds and leaflets usually appear on the edge of a plot. Symptoms are worsened by exposure to sunlight: this particularity is therefore an excellent deficiency indicator, as portions of leaflets that are less exposed are greener than the rest of the lamina (photo 3). Symptoms first occur on the lower fronds before a significant effect on yields is seen. Deficient palms often occur in patches that coincide with sandy or stony soils with low mineral reserves. For all these reasons, a corrective fertilizer application can only be exceptional and on small areas if the diagnosis is based solely on observing symptoms.

Photo 3. Two views (3A and 3B) of the same frond affected by a severe magnesium (Mg) deficiency

Photo 3B: The leaflets of the upper layer have been removed between the two points indicated by the arrows, revealing the leaflets of the lower layer that are much greener. This difference in tissue colour between low or high exposure to light is typical of magnesium deficiency.

Boron (B) deficiency

Boron (B) deficiency is difficult to detect. Experienced agronomists attribute several symptoms to this deficiency, which disrupts the terminal buds of the palms with, in increasing order of intensity: crinkling of distal leaflets, leaflet malformations (Hook Leaf) (photo 4) stunting of leaf rachises, resulting in a stump in serious cases. These are associated with the more frequent appearance of pale yellow stripes parallel to the leaflet midribs. All these symptoms sometimes indicate a true nutritional problem, especially at the susceptible age between 2 and 5 years. The most decisive criterion is the appearance of new fronds that are shorter than the previous ones, giving a flat top growth habit to young palms. However, malformations of distal leaflets, or the existence of yellow stripes, are not alone a reliable criterion for deciding on the need to fertilize, especially in mature palms. Leaflet malformations may have a physical origin at the time the frond opened, as shown by observations on $O \times G$ hybrid material in the field or in the nursery. Indeed, such malformations often occur on only one side of the frond (always the same), which suggests they are linked to the phyllotaxis of the oil palm. There may be other reasons for such malformations, notably insect damage.

Photo 4. Hook leaf of distal leaflet tips caused by a boron deficiency

Copper (Cu) deficiency

Copper (Cu) deficiency has been reported on sandy soils and peat soils. Up to three years old it can lead to some plant losses if no copper is applied. Beyond that age, the risk of copper deficiency disappears.

The first copper deficiency symptoms involve plasmolysis of young frond rachises: the palm takes on a typical arched and sagging shape (photo 5). Leaflet tips are then affected by a typical graduated brown-yellow-green discoloration. With a severe deficiency, new fronds are increasingly stunted and the deficiency sometimes kills the palm.

Manganese (Mn) deficiency

Manganese (Mn) deficiency is rare. It occurs exceptionally in situations where calcium blocks manganese uptake if it is overabundant in the soil. Deficiency is then reflected in the deterioration of chlorophyll tissue and frond stunting; these symptoms are easy to identify (photo 6). Palms displaying such symptoms require manganese sulphate fertilizer to resume normal growth and yields.

Other nutrients

Some nutrients can be deficient and limit yields though their deficiency is difficult to detect, such as phosphorus (P), or even invisible to the naked eye, as for chlorine (Cl). Yet, some P and Cl deficiencies can severely reduce yields.

Conclusion

Observing deficiency symptoms does not provide precise information on the need to adjust fertilization.

In the immature phase and the subsequent three years (i.e., up to 6 years old), when there are deficient palms it is generally possible to identify the nutrient responsible and assess the extent of the areas involved. However, the same importance is not accorded to the diagnosis if just a few palms are affected, or several hundred. As it is necessary to move quickly when palms are young (lower mineral reserves, poorly developed roots), corrective fertilization can be provided on all or part of the plots depending on the nutrients and depending on the extent of the deficiency.

In crops over 6 years old, it is not possible to rely on the existence of symptoms to make a fertilizer recommendation, firstly because nutrition may be deficient (particularly N and K) without any abnormality being detected, and secondly because there is not always a direct relation between the appearance of symptoms and a drop in yields, as already seen above for magnesium.

It is leaf analyses that can confirm the existence of a deficiency in the field and also help to back up the decision to fertilize, even when no symptoms are visible.

Analysing leaf samples to establish a diagnosis

The diagnostic tool known as leaf analysis reveals the composition of chlorophyll tissues in leaflets (other tissues are sometimes used). The diagnosis is established by comparing the content results obtained to reference values.

Photo 5. Copper deficiency on peat soil in Indonesia: arched and sagging shape of the palm

Photo 6. Manganese deficiency on a young palm planted on a karstic relief in Guatemala.

Desiccation of new fronds may kill young palms if no manganese is provided in soluble form. Once the deficiency occurs, it is easy to mark out the sectors that need treatment.

Leaflet sampling has been standardized and the practices are well shared by agronomists and scientists:

– the frond to be sampled is rank 17 on the day of sampling. Rank 9 was used at one time to establish an initial leaf analysis at 3 years. As cultural techniques have evolved, and it is now possible to have healthy rank 17 fronds at as early as 30 months, it is now recommended that this rank be used exclusively,

– the leaflets sampled lie close to the middle of the rachis. This point is easy to determine visually, even from the ground, and the composition of the leaflets seems to be most stable in the central section of the lamina. Some authors also recommend "point B" which corresponds to a change in shape of the rachis cross-section (figure 4), but it is located on the upper side of the frond and is therefore difficult to identify on tall palms.

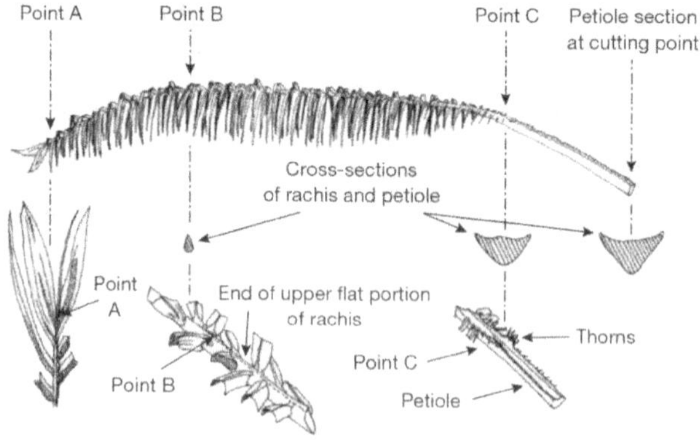

Figure 4. Oil palm frond structure

The longest leaflets making up the leaf sample are taken at mid-distance between point A and point C. A section of lamina measuring around fifteen centimetres is cut from the central part of a few leaflets and the sample is oven dried at 70°C then sent to the laboratory for analysis.

Leaflet samples are taken from either side of the rachis. Two leaflets are enough for O×G hybrids. For *guineensis* material, two classes of leaflets (underside and upper side) need to be represented, i.e., at least four leaflets will be sampled per palm. These precautions are important, as the magnesium content of leaflets depends on their position and their exposure to sunlight (Webb *et al.*, 2009).

Between 25 and 35 palms are used to make up the composite sample for analysis; they are ID-marked in the field and will be sampled each year. The size of the sample mainly varies due to sanitary risks that affect individual palms, which gradually decrease the initial population.

The position of the sampled palms inside the cultivated plot is paramount: this point is dealt with in detail in the "Choosing the palms of the reference sample inside the LSU" section, page 41.

As palms grow older, the rank 17 frond becomes increasingly high up and difficult to reach with a hooked knife (especially point B). Consequently, it happens that the entire frond is cut to take leaflet samples. In this case, sampling is no longer representative of a normally managed plantation, as each sampling operation amounts to removing 4 to 5% of the fronds produced annually, thereby reducing the leaf area accordingly. Strictly speaking, this particular treatment can distort the diagnosis deduced from the sample analysis. As an exception, such sampling involving the removal of an entire frond is only acceptable in the final years before replanting.

For more information on taking samples and preparing them for analysis, see IRHO Advice Notes (CIRAD-IRHO, 1977).

Ascertaining variability in leaf nutrient contents

Nitrogen (N)

Leaflet N contents vary between 2 and 3% of dry matter (DM) weight; it is common to read that a "normal" N content is 2.5% for *E. guineensis*. However, that value makes no sense without reference to the age of the palm because, for the same leaf rank, contents decrease rapidly in the early years from 3 years old onwards (figure 5). That decrease, which is a common occurrence in growing plant covers, is attributed to a law of dilution due to an increase in biomass. In the oil palm the age effect lessens from 12 years onwards when the fronds reach a stable length and biomass, then becomes insubstantial after 20 years.

Models

Various models have been developed based on observation data to describe variations in N content depending on age. They produce reference values to assist fertilization management (box 3). When measured contents reach or exceed the model values, no improvement in yields can usually be expected. When contents are below the model values it is determined, for each site, up to what value the deficit is acceptable.

Box 3. Models used to describe variations in nitrogen content depending on age

Indonesian model: N Ind

Tampubolon *et al.* (1989) proposed a so-called Indonesian model "N Ind" based on trial results from North Sumatra (see also the example in Colombia, figure 5). The equation is that of a parabola:
N Ind = 3.192 – 0.059 × n + 0.001 × n2
N Ind is the N content (% DM) calculated by the Indonesian model, and n the age in years.

Latin American model: N LA

In Latin America, the so-called "N LA" model also gives good results:
N LA = 2.33 + 0.7054 exp(-0.0975 × n)
N LA is the N content (% DM) calculated by the Latin American model, and n is the age in years. The N LA model can also be used for O × G hybrids (figure 6). In this case, the constant parameter is adjusted, decreasing from 2.33% to 2% DM, giving the following equation:
N LA O × G = 2.0 + 0.7054 exp(-0.0975 × n)
N LA O × G is the N content (% DM) calculated by the Latin American model for O × G hybrids and n is the age in years.

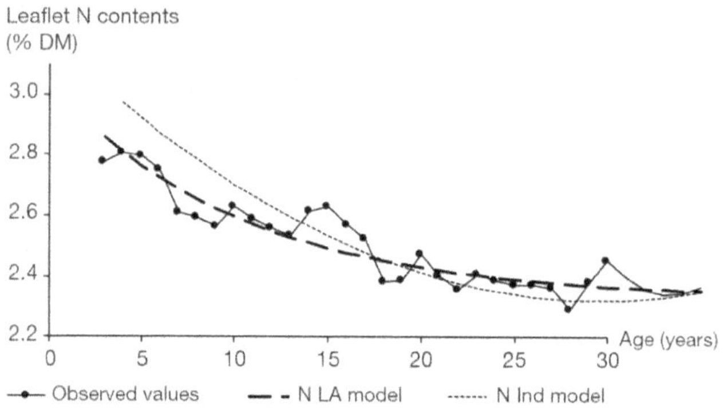

Figure 5. Variations in leaflet N contents for *E. guineensis* over a complete cropping cycle in a Colombian plantation and comparison with the N Ind and N LA models

> The mean values were obtained from 4 to 10 observed values for different planting years at a given age to smooth the fertilization effect. Contents decreased rapidly up to 12 years then moderately thereafter. This trend depending on age can be represented by the two models mentioned, N Ind and N LA. Deviations from the predicted contents were therefore due to factors other than age, notably climatic conditions and biomass production cycles.

Fertilization trial results need to be interpreted in line with a model to bring out the share of variation due to palm ageing and that due to fertilization.

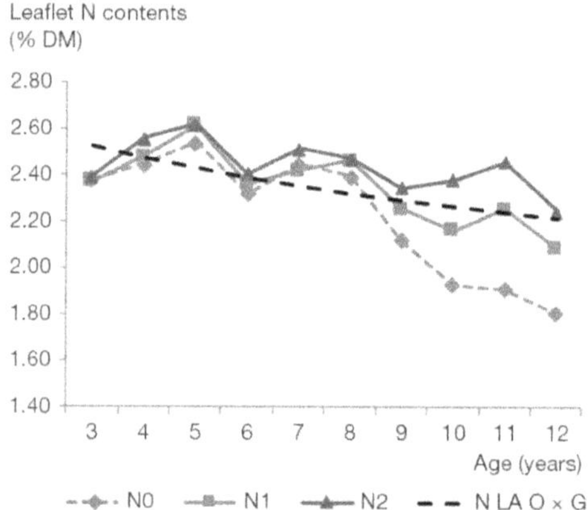

NO no fertilizer ; N1, 1.5 kg urea/palm/year ; N2, 3 kg urea/palm/year

Figure 6. Results of a trial with three nitrogen application treatments and comparison with the N LA O×G model

> In this trial with O×G hybrid crosses, the contents decreased up to 8 years for all three fertilization treatments, while remaining higher than the values of the N LA O×G model. It was only after 9 years that the nitrogen nutrition of the NO control deviated from it. The trial showed that, as of that age, the average yields recorded for NO became statistically lower than those of treatments N1 and N2. It was thus possible to confirm that the model equation predicted a nitrogen content reflecting satisfactory nutritional status at each age.

How frond rank affects contents

The rank of the sampled frond also affects the leaf N content (figure 7). As sampling errors can often occur it is essential that samples be taken by people with a perfect command of the operation.

To compare contents two years running the rank of the sampled fronds must be strictly respected, in particular using rank 17 right from the first sampling operation at 3 years old.

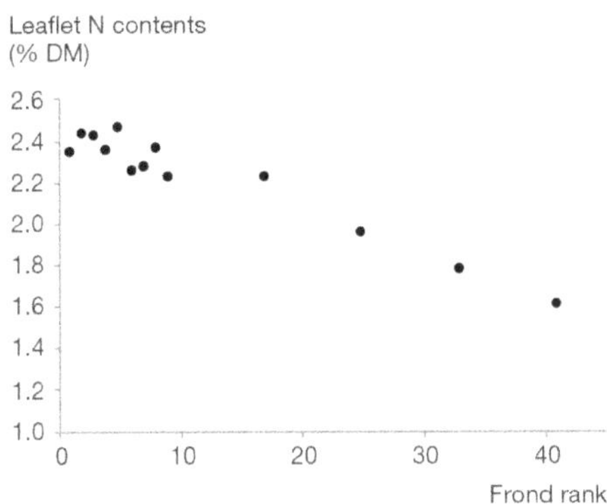

Figure 7. N contents depending on frond rank for cross LM2T × DA10D at the La Mé station in Côte d'Ivoire

An average drop of 0.02% DM is found per frond rank.

How sunlight affects contents

Although no in-depth studies have been undertaken, it seems that annual exposure to sunlight also influences nitrogen contents. This phenomenon has been reported in Ecuador where the very low sunshine exposure observed in the Quinindé region (around 900 sunshine hours/year) would seem to explain the systematically lower contents than in other locations, at the same age with equivalent fertilization and for the same planting material of *E. guineensis* origin (figure 8).

Interannual variations

N contents often vary from one year to the next over a section of the plantation within the same age range (figure 9), or even an entire plantation. Overall N contents decrease or increase from one year to another without any change in fertilization.

When interannual variations occurred in a fertilization trial studying nitrogen (figure 10), they were found to be independent of the amounts of fertilizer applied and their amplitude could be greater than the effect of nitrogen fertilizer applications.

These phenomena found on an entire plantation scale do not therefore seem to be linked to fertilizer applications, but they might be explained by physiological mechanisms that regulate N allocation to leaflets.

When interannual variations are greater than the fertilizer effect, it takes several years of observations to determine an optimum ratio compared to the N LA model.

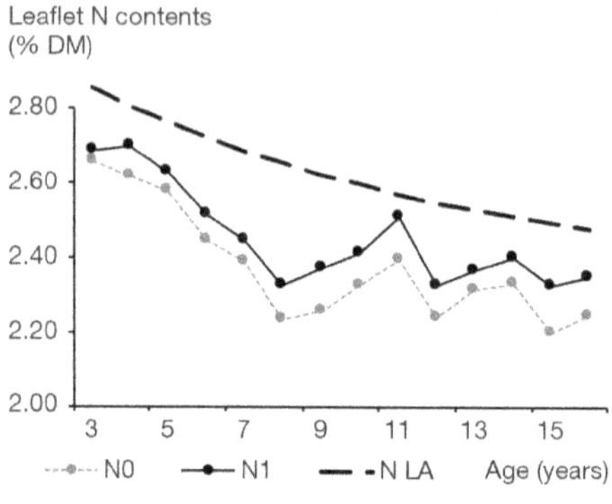

Figure 8. Effect of age and nitrogen fertilization on the leaf contents of palms in the Quinindé region of Ecuador

The average values obtained for each treatment fell steadily with age. The difference between N1 and N0 increased and became significant from 8 years onwards, but it was not possible to achieve the values predicted by the N LA model.

On average, N1/N LA = 93% and N0/N LA = 90%.

The urea applications had no positive effect on yields, and it was concluded that it was enough to achieve 90% of the model for nitrogen nutrition to be satisfactory in this plantation with low annual sunshine.

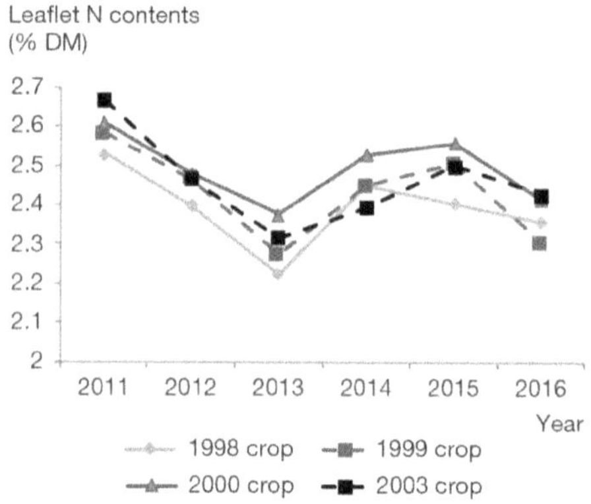

Figure 9. Example of synchronism in average N content variations observed for crops planted from 1998 to 2003 in Colombia

A notable drop in contents was observed in 2013 for the four planting years.

Conclusion

To conclude, for nitrogen nutrition management it is preferable not to use observed contents directly, but to take into account the age of the crops and determine reference values compared to a model. When contents reach or exceed the model values, no improvement in yields can usually be expected. When the contents are lower than the model it will be determined, for each site, up to what value that deficit is acceptable.

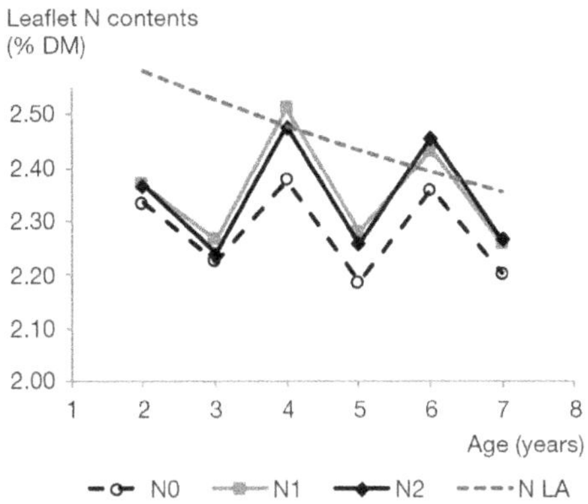

Figure 10. Effects of fertilizer application treatments and comparison of those effects with interannual variations

In this trial, the N1 and N2 urea applications significantly increased contents compared to the N0 control without fertilization. However, the interannual variations in average N contents were largely greater than the fertilizer effect.

Phosphorus (P)

Leaflet P contents vary from 0.12 to 0.19% of dry matter (DM) weight. Fertilization trials indicate that leaf contents usually respond well to P applications and sometimes lead to an increase in yields. However, such responses cannot be used to determine a threshold value indicative of a P deficiency.

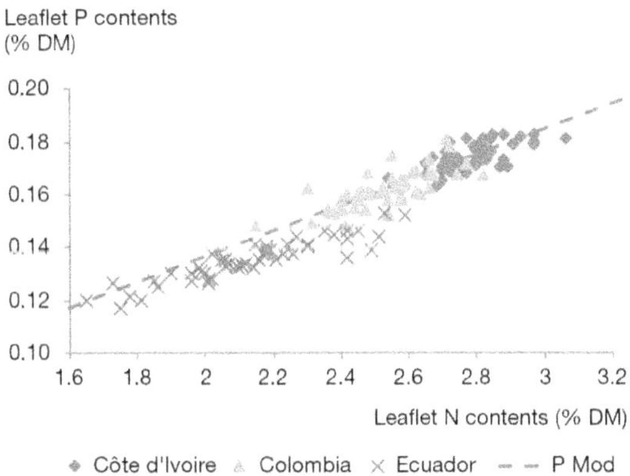

Figure 11. N and P contents (% DM) for mature *guineensis* plantations located in Colombia, Côte d'Ivoire and Ecuador

In Ecuador, N contents limited by sunshine explain low P contents. Conversely, in Côte d'Ivoire environmental conditions mean that N nutrition is abundant, which affects P contents. However, the model clearly explains the relation between N and P in each situation. The hypothesis adopted to explain the reliability of the equation is that the composition of lamina proteins varies little.

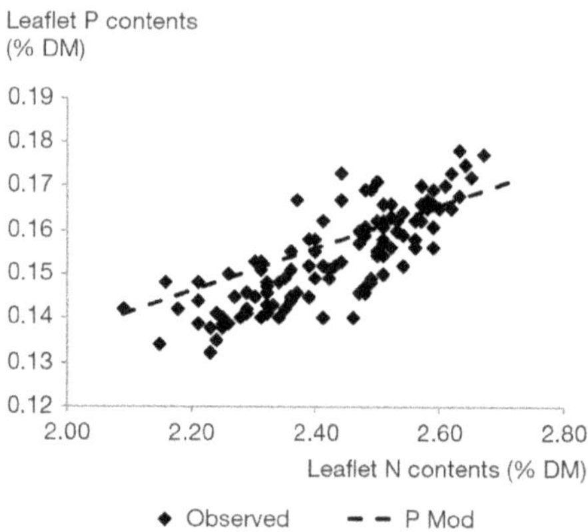

Figure 12. Relation between N and P contents for O×G hybrid material

In this plantation in Ecuador the N contents observed for 5 to 17-year-old palms were distributed along a gradient from 2.1% to 2.7% DM. The model explains most of the variation in P contents from 0.13 to 0.18% DM. The leaflet samples with the least P had a 12% deficit compared to the value predicted by the model.

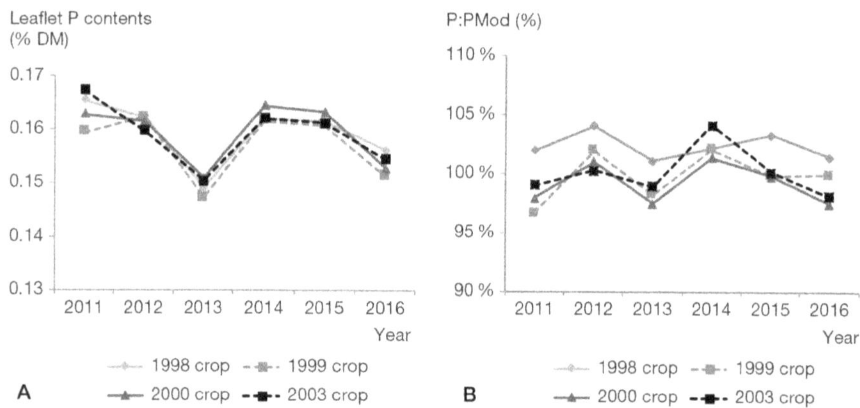

Figure 13. Interannual variation in P contents for crops planted from 1998 to 2003 (figure 13A) and its consequences for the P content:PMod ratio (Observed P:Predicted P, figure 13B)

In the example in figure 9, some uncontrolled factors caused N contents to drop in 2013 and that led to an accompanying variation in P contents (figure 13A), but the variations in the P:PMod ratio were mitigated (figure 13B). Consequently, it could not be concluded that there was a deterioration in phosphorus nutrition in 2013 for the four crop years examined. The range in interannual variations for the P:PMod ratio was mitigated compared to the variations in P contents.

P and N balance model

Use of the P and N balance model established by Ollagnier and Ochs in 1981 is recommended:

$$Pmod = 0.0487 \times N + 0.039$$

Pmod is the P content (% DM) predicted according to the N content (% DM).

This model reflects a very stable relation confirmed in several African, Asian and American countries (figure 11). On the American continent it is also applicable to $O \times G$ hybrids (figure 12).

When the P:Pmod ratio (ratio between the measured P content and the P content predicted by the Pmod model) approaches 100%, fertilizer applications have little effect on yields. If P:Pmod is below 90%, phosphate applications to correct the leaf contents will probably be followed by improved yields.

This relation between N and P is important, as it can be used to deal with factors that will cause N nutrition to vary and thereby affect P contents (figure 13).

Box 4. Interpreting the combined effects of nitrogen and phosphorus fertilizers on N and P contents

An example is provided by the results of a factorial analysis (figure 14) testing P applications in triple superphosphate (TSP) form, and N in urea form:
— P0 and N0 are the treatments without nitrogen or phosphorus fertilizer,
— P1, 0.75 kg/palm/year; P2, 1.5 kg/palm/year,
— N1, 1.5 kg/palm/year; N2, 3 kg/palm/year.
The trial very quickly showed that yields were limited by a severe P deficiency, which was confirmed by soil analyses. The first TSP rate (P1) was enough to achieve the highest yield.

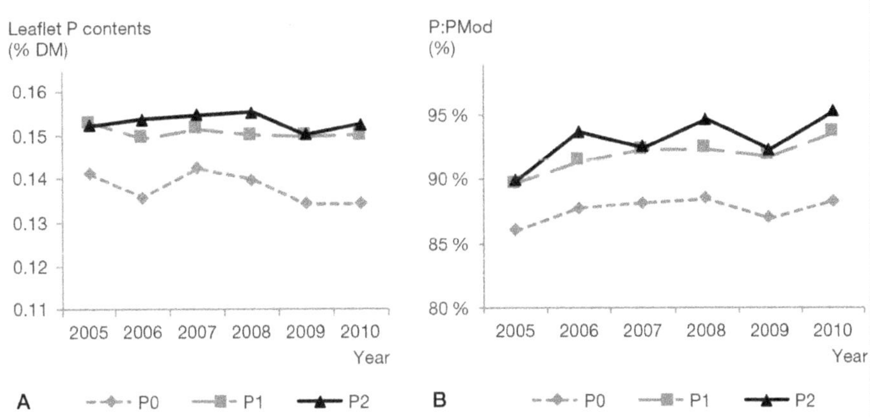

Figure 14. Responses of leaf P contents (figure 14A) and of the P:PMod ratio (figure 14B) to TSP applications

The effect of TSP on phosphorous nutrition is shown in figure 14, revealing significant increases in P contents (figure 14A) and in the P:PMod ratio (figure 14B) obtained with applications P1 and P2. The increase in yield can therefore be explained by better phosphorus nutrition.

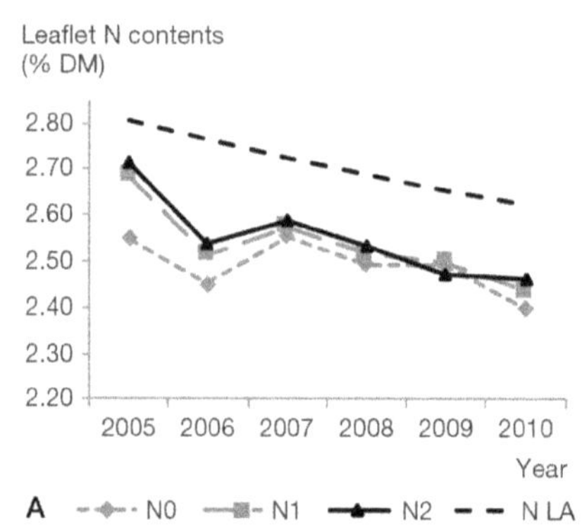
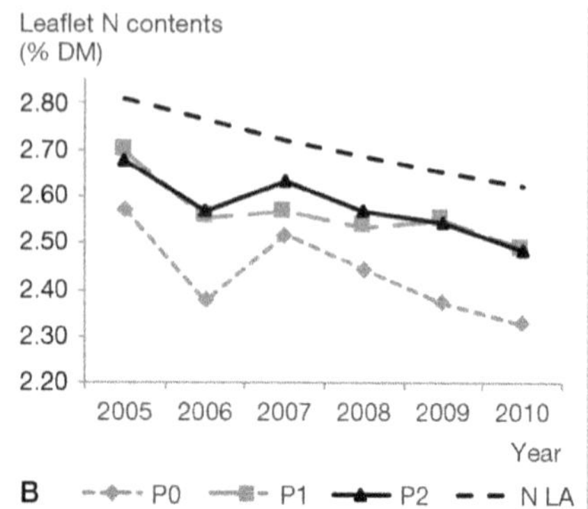

Figure 15. Responses of leaf N contents to urea applications (figure 15A) and to TSP applications (figure 15B). N LA is the value predicted by the model, which describes the variation in N content depending on palm age

Urea applications did not have any effect on yields, which reflected the absence of response in leaf N contents. A urea effect was seen on N contents in 2005 at 4 years, but it then disappeared, even though the contents were well below the values of the N LA model. However, to decide about nitrogen nutrition it is necessary to consider figure 15B, which shows that N contents increased significantly with the TSP applications. This fertilizer reduced the deviation from the N LA model.

All in all, on these soils containing very little P, TSP applications improved both P nutrition and N nutrition. Given these results, it can be recommended that the P:PMod ratio should reach 90%. They also confirmed that nitrogen nutrition was satisfactory with an N:N LA ratio of 95% or over, obtained after correcting the P deficiency.

Potassium (K)

Potassium is the key element in oil palm nutrition, and it has played a major role since 1950 as a tool in steering the development of leaf analysis (LA). However, specific norms have to be adapted for each situation: defining a range of optimum potassium (K) contents is the main difficulty encountered by agronomists in charge of oil palm fertilization.

It is necessary to determine the optimum leaf K contents in each situation. In the original work (Ollagnier and Ochs, 1981) leaf K contents were often considered optimum when they exceeded 1% of dry matter (DM) weight. However, it was subsequently shown that fertilizer savings could be made by adjusting optimum contents to each situation, in order to take the type of soil, type of climate and potential yields into account. Optimum contents of between 0.80% and 1.0% DM were then published for *E. guineensis* depending on environmental conditions (Caliman *et al.*, 1994). For Goh and Härdter (2003), "normal" contents lie between 0.90 and 1.30% for most soils.

The climate, the planting material and the soil are factors independent of fertilization that also influence optimum leaf K contents.

Influence of the climate

A more or less marked dry season has consequences for leaf K contents. For Ollagnier *et al.* (1987) quoted by Caliman *et al.* (1994), the optimum K content passes through a maximum for mean annual water deficits of around 200 mm (figure 16). This relation is primarily based on data gathered in Africa and is doubtless not applicable to other continents.

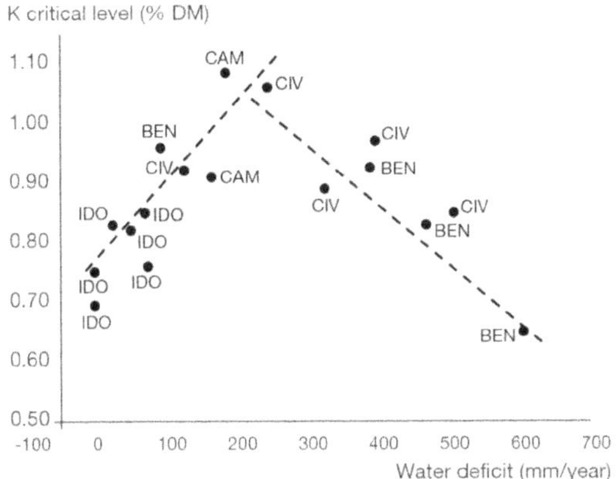

BEN: Benin, CAM: Cameroon, CIV: Côte d'Ivoire and IDO: Indonesia

Figure 16. Relation between the optimum K content (critical level) and the annual water deficit in Africa and Indonesia (according to Caliman *et al.*, 1994)

Influence of the planting material

The genetic origin of the planting material is also an important parameter (Tan and Rajaratnam, 1978; Jacquemard *et al.*, 2009). The main problem when taking this factor into account is currently the difficulty in establishing or predicting optimum contents specific to each genetic origin. It is possible to detect content differences experimentally for crosses receiving the same fertilization (figure 17), but their consequences for nutrient requirements remain unknown.

Influence of the soil type

An abundance of exchangeable calcium (Ca) in the soil tends to lower leaf K contents. The phenomenon seen when chlorides are applied is explained by preferential allocation of calcium to leaflets, to the detriment of potassium. The mechanism is described in the "Taking into account soil calcium contents when using KCl" section, page 61. It is worth noting that on calcium-rich soils KCl applications reduce leaf K contents up to around 0.80% DM. Such contents are low compared to the reference values, but these deficiencies usually have no impact on yields.

Calcium (Ca)

In mature plantations at least 12 years old leaf calcium contents are almost always between 0.50% and 0.80% of dry matter (DM) weight. They may approach 1% when this nutrient is abundantly present in the soil.

Observing leaf Ca contents helps in understanding the dysfunctioning observed with other cations, especially K and sometimes Mg. Ca applications (application

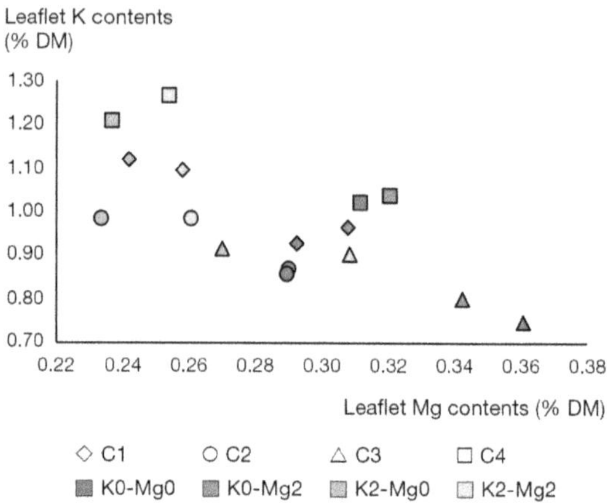

This mineral nutrition trial compared 3 KCl rates (K0, K1, K2) and 3 kieserite rates (Mg0, Mg1, Mg2), and studied four crosses (C1, C2, C3, C4) known for their contrasting leaf K and Mg contents. The effect of extreme rates (0, no fertilizer; 1, "intermediate" rate for KCl and kieserite; 2: maximum rate for KCl or kieserite) was examined to test the crosses after four years of protocol application.

Figure 17. Planting material specificity test

> The results confirmed the specificity of the crosses: C3 and C4 showed high leaf contents for Mg and K respectively, C2 displayed low Mg and K contents (according to Dassou *et al.*, 2018).

of phosphates, ameliorators with lime or gypsum) can reduce K and Mg contents. It is also worth analysing the relation between Ca and Cl contents in experimental designs, as explained in the "Taking into account soil calcium contents when using KCl" section, page 61.

However, it is impossible to determine an optimum leaf Ca content, as there are no known deficiency symptoms. Neither are there any yield responses to calcium fertilizer application, although liming carried out to improve soil structure may result in better yields when it leads to a better water supply.

Magnesium (Mg)

Leaf Mg contents range from under 0.10% of dry matter (DM) weight, in cases of a severe deficiency, to between 0.30 and 0.40% DM when the soil has good reserves. In a properly controlled fertilization trial a significant drop in yields was found starting from a concentration of under 0.16% (Dubos *et al.*, 1999), whilst leaf symptoms started from a concentration of 0.20%.

A magnesium deficiency is often more spectacular than serious. For Webb *et al.* (2009), magnesium is rapidly translocated from the oldest tissues, where symptoms appear first and become the most intense, to younger tissues. Symptoms are the most intense in parts of the foliage exposed to sunlight, as reported in the "Can deficiency symptoms be trusted to recommend fertilizer applications?" section, page 16. Consequently, palms planted on the edge of a plot seem deficient more often.

Site effect

A leaf Mg content of 0.24% DM or more has often been considered satisfactory. It is nonetheless likely that this threshold is overestimated (Ollagnier and Ochs, 1981) compared to the impact of this nutrient on yields.

In Peru, despite the existence of deficiency symptoms, contents of between 0.18 and 0.22% DM were considered enough as yields did not rise with higher contents.

In the Quinindé region of Ecuador, where exposure to sunlight is low, both leaf Mg and N contents are lower than at other sites with the same planting material and equivalent fertilization (figure 18). In that region, Mg contents of between 0.16% and 0.20% M are enough for *E. guineensis* material of Deli × La Mé origin.

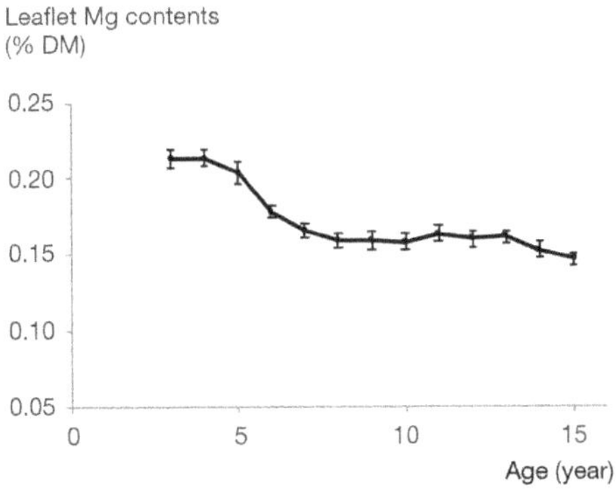

Figure 18. Average leaf Mg contents depending on age in the Quinindé region (Ecuador), where exposure to sunlight is low

> The drop observed between 3 and 7 years was due to an increase in leaf biomass. It was similar to that observed in other circumstances, but the mean contents become stable at around 0.16% DM, whereas the fertilizer applications ought to have kept them above 0.20% DM. The fertilization trials confirmed that neither leaf contents nor yields increased significantly with the magnesium fertilizer applications and that this lack of response occurred for the different types of soil in the plantation.

Planting material origin

The origin of a planting material affects leaf Mg contents. For instance, *E. guineensis* Deli × Yangambi material displays Mg contents around 0.04% DM higher than *E. guineensis* Deli × La Mé material (F. Corrado, pers. comm.). Advances are expected with the results from some specific experimental designs set up by CIRAD and its partners to study how the type of planting material affects leaf K and Mg contents.

In Latin America, where O × G hybrid material is now very widely used for phytosanitary reasons, there is substantial variability from one cross to another. These differences occur in the foliage colour seen in the field (photo 7), but also in different leaf Mg contents with identical fertilization (figure 19). It is sometimes tricky to interpret analysis results, but it is hoped that further work will help to specify norms based on origins.

Photo 7. Specificity of the genetic origin of O×G material for magnesium nutrition: comparison of two hybrid crosses of Coari×La Mé origin planted side by side in the same plot and having received the same fertilization up to 4 years old

The palms in the right-hand row (progeny TT3314) display intense symptoms, while those in the left-hand row (progeny TT3306) are free of yellowing, showing the specificity of the genetic origin of O×G material for magnesium nutrition.

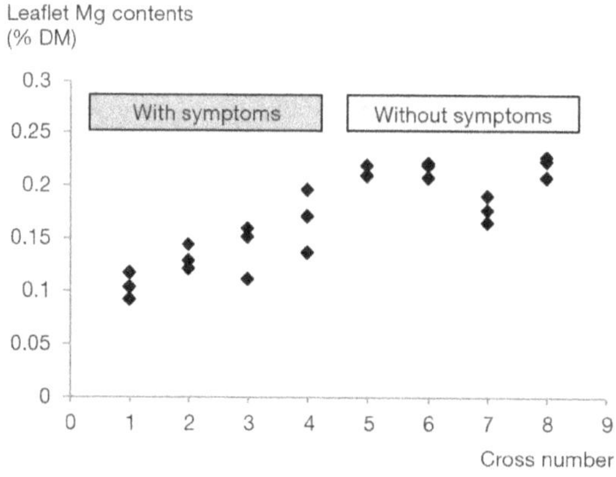

Figure 19. Effect of planting material on magnesium nutrition (crosses 1 to 4 with the same *E. guineensis* pollen in common, crosses 5 to 8 derived from another pollen). Trial conducted in Ecuador

At 5 years old, crosses 1 to 4 displayed Mg contents lower than those of crosses 5 to 8. The symptoms were also more pronounced in the first group.

Chlorine (Cl)

Leaf chlorine (Cl) contents are usually satisfactory as potassium chloride (KCl) is routinely used in fertilization to cover K requirements.

Cl deficiency occurs in areas far from oceans where atmospheric supplies do not exist. Such is the case of the Amazon basin (Peru, Ecuador, Brazil, Colombia), but also in the Magdalena Valley in Colombia, where the deficiency was discovered by Ollagnier and Ochs (1971).

Leaf contents are satisfactory as of 0.50% of dry matter (DM) weight. This threshold is quickly reached at the start of yields with KCl applications. In mature crops the applications needed to satisfy K requirements are always enough to satisfy Cl requirements at the same time and leaf Cl contents are between 0.60 and 1% DM.

An annual chlorine analysis is recommended if the plantation is located in a soil deficiency zone. In other cases occasional checks are useful at the start of production at 3 and 4 years.

In fertilization trials using KCl, chlorine also needs to be analysed for a satisfactory interpretation of KCl effects on yields.

Sulphur (S)

It is rare to find proven cases of sulphur deficiency responsible for a drop in yields. Leaf S and N contents are closely correlated and the symptoms described in young crops are similar to those of a nitrogen deficiency.

Ollagnier and Ochs (1972) considered the critical level for sulphur contents to be between 0.20 and 0.23% of dry matter (DM) weight. More recently, Gerendás *et al.* (2009) lowered that critical level to 0.15%.

Sulphur leaf analysis is not standard practice when checking the nutrition of oil palm plantations, no doubt due to the additional cost and the limited effect of this nutrient on productivity. In order to ensure a supply of sulphur it is recommended that at least one fertilizer is a sulphate, such as ammonium sulphate [$(NH_4)2SO_4$] used for N requirements, or kieserite (totally soluble magnesium sulphate) for Mg requirements.

Trace elements: boron, copper, iron, manganese

Case of boron (B)

Of the so-called trace elements, boron (B) is the one most routinely analysed and is included in fertilizer applications. The diagnosis of a boron deficiency and management of applications remain controversial for different reasons.

Deficiency symptoms are linked to a dysfunctioning of the terminal bud of palms resulting in more or less intense malfunctions in the leaf tissues produced (see page 19). Actual deficiencies are more frequently observed between 2 and 5 years old and substantially reduce the leaf area. This leads to a drop in photosynthetic activity and in productivity, which are effectively corrected by borax applications. True deficiencies are rare in the mature phase except on certain soils that are intrinsically very impoverished, as encountered in Amazonia. When the leaf area is not reduced, white stripe symptoms or malformed distal leaflets do not indicate the actual existence of a deficiency, as such symptoms may have other origins; they can even be more frequent when fertilizer is applied (figure 20). Concluding on a boron deficiency by observing isolated symptoms in mature crops calls for considerable caution.

Analysing boron contents in ppm of the dry matter weight of frond 17 does not provide any useful information. Firstly, B determination in the laboratory lacks precision with a confidence interval of ± 2 to 3 ppm. Secondly, B contents vary from 10 to 30 ppm, or even up to 50 ppm, shortly after an application. It is often read that they need to be over 12 ppm, but there is no experimental result to prove that this value corresponds to a deficiency threshold. No reduction in leaf area is found for contents under 12 ppm. On the other hand, young palms can be found with short fronds when the frond 17 contents are satisfactory. This absence of relation between vegetative status and leaf contents is due to the fact that, unlike other nutrients, boron is a particularly immobile element. Frond rank 17 is probably not suitable for observing a deficiency at a given moment as it will have been emitted 6 to 8 months before the date on which the analysis results become available. A high content does not therefore reflect a reserve potentially recyclable to newly opening fronds in which the deficiency will be expressed.

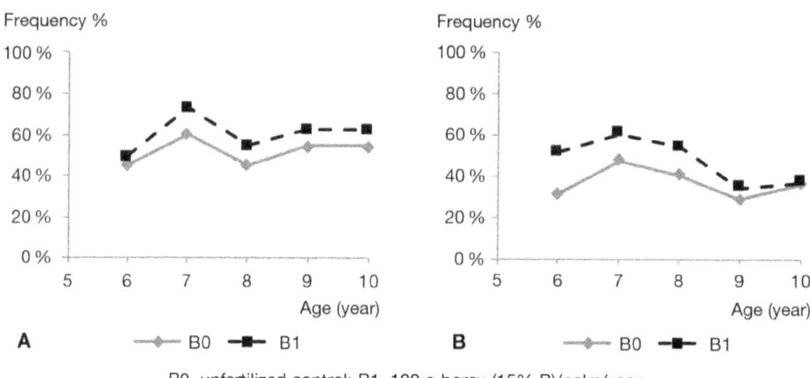

Figure 20. Boron fertilization trial between 6 and 10 years in Côte d'Ivoire: percentage of palms with white stripes (figure 20A) and distal leaflet malformations (figure 20B)

> From 7 to 10 years, the existence of white stripes was significantly more frequent with borax application, without the effect being explained. Up to 8 years, leaflet lamina malformations were more frequent for the palms receiving boron, albeit not statistically significant. Borate applications did not have any effect on the yields observed, which confirmed that there was no deficiency and the symptoms displayed were not relevant.

Case of copper (Cu)

The critical level for copper is 3 ppm; above that level there is virtually no risk of a deficiency. However, deficiencies can occur below 2 ppm. As copper analysis accuracy is around 1 ppm, it is difficult to use the laboratory result threshold to decide whether or not to apply copper sulphate. It therefore remains necessary to monitor symptoms in the field. On peat, applying copper sulphate ($CuSO_4$) to the soil surface is an effective preventive measure in blocks with a proven copper deficiency. Four applications are staggered from planting up to 18 months, and it is rare for correction to be needed after three years. Copper sulphate can also be used as a corrective measure, splitting applications to prevent immobilization of the product in organic complexes.

Case of iron (Fe)

Iron (Fe) deficiency exists on peat but is very temporary: it occurs in the first year after planting, it is usually non-lethal and the palms recover a normal vegetation without any need for correction. Iron deficiency is characterized by alternating green and yellow stripes along the lamina (interveinal chlorosis).

Case of manganese (Mn)

Leaflet manganese (Mn) contents vary widely between 100 and 600 ppm, without it being possible to establish a critical level with any certainty when palms do not display a deficiency. Affected palms can be treated individually by manganese sulphate applications until the foliage has recovered.

Interpreting leaf contents considering the specific characteristics of a plantation

Leaf analysis results need to provide a diagnosis of nutritional status at a given moment in time (deficiency, satisfactory content, very high content), from which to deduce fertilizer recommendations (corrective or maintenance rate, halting of applications). The above examples show how essential it is to know the context in which results are obtained. A given leaf content will be considered satisfactory and will only need slight adjustment or, on the contrary, will call for greater correction depending on the age of the palm, the type of planting material, soil properties and the climate.

In addition, the leaf analysis laboratory may be responsible for some of the variability found in the results obtained (and the soil analysis laboratory for CEC and exchangeable cations), depending on its instruments, analysis methods (several methods for N and P, several calcination methods, etc.) and how it operates (frequency and precision of self-inspection protocols, certification, inter-laboratory analyses). It is essential to choose a laboratory capable of providing high-quality mineral analyses with a stable degree of precision over time.

To conclude, interpreting leaf mineral nutrient contents is tricky, calls for experience and needs to take into account the specific characteristics of each plantation, because the latter will affect the optimum reference contents. It is therefore recommended that leaf analyses be combined with one or more fertilization trials that will help to improve result interpretation based on fertilizer applications and yield responses. This in situ experimental procedure is explained on page 47 ("Fertilization trial principles") and calls for several years of investigation.

2

Plantation sampling for ongoing mineral nutrition monitoring

Leaf contents react not only to fertilizer applications, but also to other factors (age, planting material, climate, soil). Before comparing them to reference values adapted to each site (see "Determining the optimum content range per nutrient", page 54), analysis results are needed periodically for all the years of planting in the plantation.

A plantation is not a uniform environment, so analyses have to be multiplied to account for spatial changes, along with the age, the planting material and soil type. For practical and economic reasons there is a limit to the number of samples that can be analysed each year for a given plantation. The plantation therefore has to be structured into small leaf sampling units that are as uniform as possible. The aim is to guarantee that the analysis results reflect the effects of the fertilizer recommendations applied and that the annual results effectively cover the whole of the plantation. The norms for creating leaf sampling units, selecting the palms that will make up the samples and the sampling techniques must be strictly adhered to, so as to ensure a precise guidance tool.

For each plantation the trials, or outside advice, will lead to values being adopted that are considered optimum for mineral nutrient contents. At the same time, a guidance tool will be deployed with a view to achieving optimum values throughout the plantation by analysing leaf samples that cover all the years of planting.

The principle consists in "dividing" the plantation into small units that are each associated with a reference leaf sample. Leaf samples are taken each year and, depending on the analysis results, fertilizer recommendations for each unit are drawn up for the following year. These units are called "Leaf Sampling Units" (LSU).

Dividing the plantation into several leaf sampling units

Leaf sampling units (LSU) are working units marked out by easily observable borders, usually tracks. The LSU comprises one or more neighbouring plots. This arrangement facilitates the checking of fertilizer applications and guarantees that all the palms in the LSU are treated in the same way. Other data are also acquired on a plot scale, such as yields and disease records, and those data can be combined for the LSU. All this information can be compared to the leaf contents (box 5).

Each LSU must comprise a uniform population of palms (planting material, planting year) cultivated under uniform conditions (previous crop cover, land preparation prior to planting, cultural techniques, soil type and topography). While this requirement can easily be controlled for genetic characteristics and cultural practices, it is not so for the soil and topography, for which heterogeneity can be detected late. In an ideal plantation, where such information would be mapped, track layouts and plot delimitations could be designed to ensure uniformity within each LSU. In practice, plantations are nearly always structured in plots respecting a regular grid pattern with parallel collection tracks aligned Nort-South and East-West. It is once that grid has been established that the LSUs inside the plots are mapped.

LSUs vary in size: the norm is usually 50 ha and the minimum is rarely under 30 ha. The average area represented by a sample is a compromise that depends on the other particularities of the plantation: the number of samples it is possible to have analysed according to a fixed calendar, variability in the soil properties and topography of the plots, along with the uniformity of agricultural operations when preparing the plots. Under very uniform soil and topography conditions some LSUs can reach or exceed 100 ha. When the soils and topography show large disparities the size of the LSUs can be reduced to a single plot (usually from 20 to 30 ha).

Box 5. Plan to feed robust databases over time

As for many tree crops data acquired in oil palm plantations are examined by grouping several years of records, up to a decade, and combining several variables available on different scales (e.g., yields per plot, leaf analyses per LSU and chemical analyses per soil unit). Over long time spans the structure of the plantation will undergo changes, especially when replanting. Data need to be organized in a way that enables rapid risk-free analysis, meaning that each object enabling the acquisition of a value at a given time must be uniquely identified and its properties must be described (metadata). This is the case for LSUs, with a unique registration number and described by the list of palms in the plots used to take the samples of the LSU. With this information, it is possible to examine the relations between the leaf contents and other variables collected on a plot scale, such as yields, the planting material, etc. All this information needs to be centralized in a database and shared, to counter any renewal of staff in charge of data gathering and analysis.

Planning the leaf sampling schedule

Leaf sampling fits in with the agricultural calendar of the plantation. For repeatability reasons, the period defined as most suitable must be respected from one year to the next. The choice of period depends on several factors (figure 21):
– availability of teams specialized in the task,
– weather conditions, as samples must be taken in a low-rainfall period,
– the mature crop production cycle. Under given climatic conditions, productivity is not evenly distributed throughout the year. When the harvesting peak occurs each year at a certain time, sampling should be programmed before or after the peak,
– the lead-time for each activity, i.e., the time taken by the laboratory to obtain analysis results, the time taken to draw up the fertilizer schedule, fertilizer purchases and their delivery. These times need to be compatible with the optimum application periods.

Each plantation needs to have its own schedule, with a great deal of thought and feedback, in order for it to be operational. The aim is to ensure that the fertilizer schedule defined after a leaf analysis campaign is implemented two months before the next samples are taken.

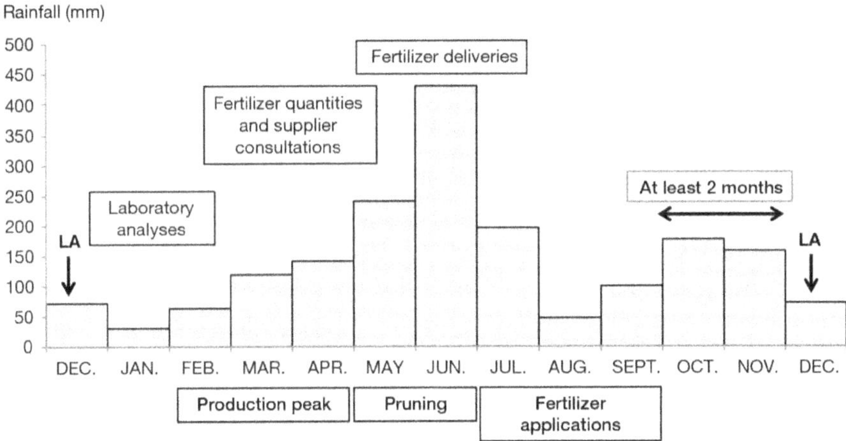

Figure 21. Example of an operational fertilizer schedule for mature plantations over a one-year cycle

> Leaf sampling (in December, noted LA in the figure) and fertilizer application (July to September) were programmed over low-rainfall periods and when labour was not greatly occupied by other essential operations, such as harvesting and pruning.

There are always unforeseen circumstances and the fertilizer application schedule may be delayed. Whatever that delay, leaf sampling must be maintained on the scheduled date, as the leaf analysis provides an annual point of reference in a standardized physiological environment, and it is important to compare nutritional status from one year to the next. It is also important not to carry over the delay into the next campaign.

Choosing the palms of the reference sample inside the LSU

The palms making up the reference leaf sample for each LSU are chosen visually. Each of them must respect vegetative development norms, be productive and be surrounded by palms with similar characteristics.

The position of the sample palms inside the LSU is the second most important criterion, as the main aim of the reference leaf analysis is to make the best fertilization decisions for each LSU with a view to optimizing its productivity (box 6).

Restricting reference sampling to a uniform section of the LSU

To make the best decision for each LSU, it is advisable to restrict reference sampling to a uniform section of the LSU that decisively contributes to the yield of the unit as a whole. Sampling is focused on one facies to obtain the most accurate recommendation possible for that part of the LSU. Several principles can help in choosing one section rather than another:

– when the LSU is fairly uniform and only a small area of it differs through its soil type (see the example in box 6), the sample should be taken from the largest section. In the example shown here, it is the section with a good Mg status (85% of the area). Thus, the section with an Mg deficiency (15% of the area) is discarded, but will undergo a special leaf analysis (see "Taking special leaf samples to check specific zones in the LSU", page 45) to determine the corrective fertilization needed to periodically control the recovery of the palms,

– when the LSU displays several facies, where the soil properties or the topography might induce variations in potential yield, the one that contributes most to the volume of bunches produced by the LSU should be chosen. Some facies will therefore be discarded, but periodical checks will be carried out to keep track of and, if necessary, correct the mineral nutrition in those zones (see "Taking special leaf samples to check specific zones in the LSU", page 45).

Box 6. A mistaken good idea: a composite sample representative of the whole LSU

Some agronomists choose to sample palms in a grid covering the entire cultivated area of the LSU taking, for example, 3 or 4 palms per row, and every 10th row. This system accurately estimates the average composition of the leaflets for the population of the LSU, but it does not account for the specific needs of the different zones of the LSU and the "average" decision that will be taken based on that sample will not be optimum.

If the LSU is heterogeneous (topography, soil chemical properties, soil depth and cultural practices), which is often the case, yield potential and the availability of nutrients will vary inside the LSU. By using an average content to decide for the whole LSU, there is a risk of recommending an inappropriate fertilizer rate for a large share of the planted area.

A simplified example is illustrated in figure 22 and table 4: it involves a plot where 85% of the area is occupied by healthy oil palms and 15% occupied by palms with a magnesium deficiency. A very simple fertilization chart varies the magnesium carbonate ($MgCO_3$) recommendation based on leaf contents (table 4): the recommendation corresponding to the average leaf content is not satisfactory for either of the two classes of palms: it is too low for the deficient palms and pointless for their healthy counterparts.

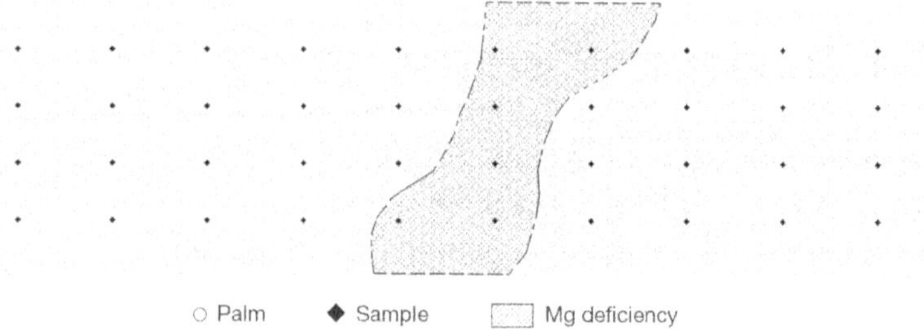

Figure 22. Systematic sampling of a heterogeneous LSU

The LSU is occupied by some highly Mg-deficient palms on 15% of the area (leaflet Mg content of 0.08% DM) and well-provided palms over the rest of the area (Mg: 0.25% DM). The average leaf content will be around 0.22% if the sample proportionally represents the two classes of palms.

Table 4. $MgCO_3$ fertilization table (kg/palm/year)

Leaf content (% DM)	0.14	0.16	0.20	0.24
$MgCO_3$ rate (kg/palm/year)	2.5	2.0	1.2	0.8	0.0

This table is used to recommend a fertilizer rate in line with the leaf Mg content. According to the table, the deficient palms (Mg: 0.08% DM) require 2.5 kg of fertilizer, while the healthy palms (Mg: 0.25% DM) do not need to be fertilized. Yet, the average content of 0.22% DM calls for an application of 0.8 kg of fertilizer for all the palms, which is totally unsuited to the reality.

This procedure is preferable to distinguishing between different facies (e.g. bottom-lands and plateaux) inside an LSU and representing them by distinct leaf samples. The main reason is that it is difficult to apply different fertilizer rates depending on the position of each palm inside a plot without making mistakes.

Illustration of unit choices between non-majority facies

We present the example of three neighbouring LSUs in a plantation with variable topography, where several arrangements are possible for the positions of the palms used for the reference leaf sample (figure 23).

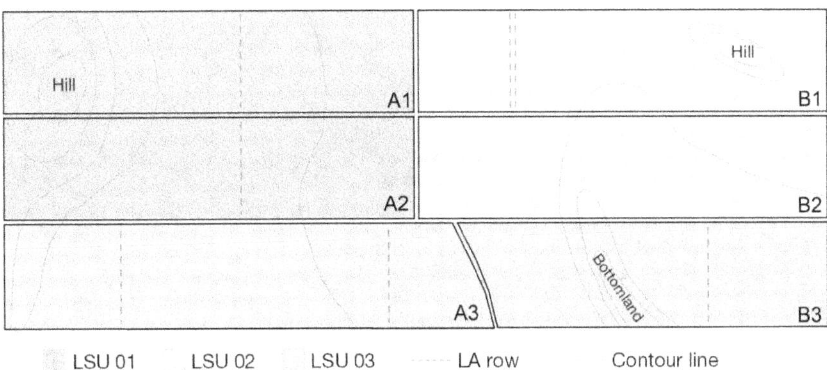

Figure 23. Examples of row choices for leaf sampling

Three leaf sampling units (LSU 01, 02 and 03) were established by respectively combining plots A1 and A2, B1 and B2, and A3 and B3. The topography comprised some small areas occupied by hills and small water courses, where the palms were not used to make up the leaf samples. Several choices were possible for the position of the rows to be used for leaf sampling. They were always located in the gently sloping zones occupying most of the plots, where sampling was focused.

For LSU 01, the intermediate topographical position (mid-slope) was chosen and the two sampling rows were located in two plots, but in a straight line to facilitate leaflet collection

for LSU 02, the two rows were side by side, to avoid the hilly relief and the head of the bottomland; this arrangement in paired rows facilitated sampling and checking of the leaf rank sampled

for LSU 03, which was flat and uniform, the sampled palms were distributed along three rows that covered the whole of the LSU and avoided the bottomlands.

When and how to select the palms used for leaf sampling

It is essential to be able to choose the positions of the palms used to take optimum fertilization decisions as soon as possible, i.e., as of 3 years old. If topography and soil maps are available, the most suitable zones will be shortlisted. For the final choice, the immature young crop stage enables detailed observations that will be decisive, especially if no other sources of information are available:

– at planting time it is fairly easy to assess and map field observations revealing zones where palm growth will be limited. By examining the planting holes hydromorphic issues can be detected, along with variations in texture, stone content, compacted surface horizons. This operation generally provides a good assessment of the uniformity of the first 30 centimetres of soil. This can be completed by an examination of the soil with an auger down to a depth of one metre,

– at around three years old, before choosing or sampling for the first reference leaf analysis, palm vigour can be assessed by a sample survey of rank 17 frond lengths. Drones also offer an excellent way of observing and verifying plot uniformity for delineating zones where growth and yields will be limited (photo 8). All the observations gathered (existence of stony or hydromorphic zones) will be mapped.

Photo 8. Aerial view of a 2-year-old plot

This view reveals heterogeneity reflected in growth differences, despite a flattish relief. The least developed young palms are probably suffering from excess water, which can affect nitrogen nutrition. This age is suitable for detecting high mortality zones and the planting material effect if several crosses are used. These aspects are taken into account when choosing the reference leaf sampling palms and also those for special sampling (see next page).

Once the leaf sampling rows have been chosen, it is essential to mark the palms with paint to ensure that the same population is always sampled. It will thus also be possible to check sampling quality (respect of frond rank, position of leaflets). The paired row design (figure 23) is ideal for observing foliage status (are there any deficiency symptoms?) and the environmental conditions that might explain the lowest leaf contents. These field checks that contribute towards the precision of decisions are not possible when sampling covers the entire LSU (figure 22).

Taking special leaf samples to check specific zones in the LSU

The reference samples have to be dedicated to a single facies corresponding to a single soil unit or a topographical position, to the exclusion of other situations existing in the sector. This method should notably be adopted when particular structures are installed on small areas, such as for erosion control (terraces, stop bunds).

For zones not taken into account in the reference leaf sampling, samples should be taken for routine tests every 2 or 3 years, up to 5 years old, to check mineral nutrition status, which is guided by the reference analysis (figure 24).

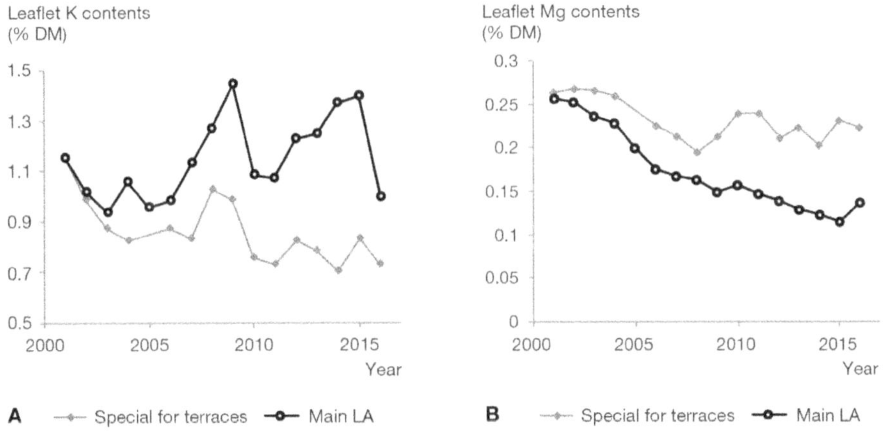

Figure 24. Example of special leaf samples to determine K fertilization (figure 24A) and Mg fertilization (figure 24B) on a minority facies (terraces) in Ecuador

In this plantation, terraces were created mechanically to plant palms located on steep slopes accounting for 15% of the area. These terraces received the same fertilization as palms in the flat zones where the main leaf sampling rows were located. A few special samples were organized to validate, or not, the decisions taken for the terraces. After several years, it appeared that additional potassium fertilizer applications (e.g., KCl) were needed on the terraces to improve K contents. On the other hand, it was found that magnesium fertilizer applications could be reduced or halted.

Special samples are also used when abnormal symptoms appear suggesting that a deficiency is becoming established in certain zones. When the responsible nutrient has not been identified with certainty (such is the case for trace elements, for which deficiencies are rare) samples need to be taken from palms affected by the symptoms and also from symptom-free control palms located nearby. Mineral nutrient contents are then compared. When it is decided to apply corrective treatment locally, samples will be taken again from the deficient palms to assess the improvement in contents.

3

Adapting the decision-support tool to local conditions: taking the specificities of each site into account

We now have a way of structuring a plantation in leaf sampling units (LSU) to obtain good quality analysis results each year.

A diagnosis is established from these analytical results by comparing them to reference contents that are fine-tuned in line with the agronomic and environmental context in each situation. Fertilization trials help to specify optimum norms for the main mineral nutrients and develop fertilizer schedules used to achieve satisfactory leaf contents everywhere on each annual check.

When there are no local reference values available to monitor palm mineral nutrition, leaf contents among the norms commonly accepted by the profession are arbitrarily used (see "Ascertaining variability in leaf nutrient contents", page 23). They are considered to be the optimum contents and fertilizer schedules that enable them to be achieved are drawn up. However, it needs to be checked retrospectively that the contents adopted as being optimum actually were so. The strictest way of testing the initial fertilization guidance values is to set up fertilization trials. This approach is based on yield and leaf content responses to fertilizer applications. It takes from 5 to 10 years to draw conclusions, but it makes it possible to determine native deficiencies that limit yields and also the range of optimum contents above which yield is no longer limited by the nutrient in question. Trials also indicate the contents it is possible to reach with the fertilizer application rates tested under trial conditions.

Fertilization trial principles

Choice of treatments and experimental designs

The aim is to test how the main mineral nutrients affect yields under the soil and climatic conditions of the plantation. The nutrients to be tested are chosen according to information already available (e.g., soil analyses that might indicate that P, K and Mg reserves are low). Potassium, the nutrient most consumed by oil palms, is virtually always provided in the form of potassium chloride (KCl). Consequently, the effect of chlorine is also tested: it has to be taken into account when interpreting results as chlorine deficiency does exist (Ollagnier and Ochs, 1971).

Trials are usually factorial designs that combine several nutrients, each applied at different rates, and they test all possible combinations. Factorial designs are appropriate as they can be used to compare the application rates of a given fertilizer with each other, with several replications for each comparison, but without having to replicate the entire experimental design, which would mean having to occupy large cultivated areas. They also allow several types of analyses, from the simplest to the most complicated, depending on the results expected from the trial.

The testing of true controls with no application of the tested nutrient (N0, P0, K0, etc.) is recommended. Such controls provide information about deficient content thresholds and about the resilience of soil reserves, as will be seen later ("Assessing soil reserves" section, page 70).

The other application rates are chosen in arithmetic progression (rate a, rate 2a, rate 3a, etc.). The extent of the chosen application rates varies depending on the age of the palms and covers a range of requirements defined according to soil type, climate and potential yields. In particular, up to 5 years old the rates applied only account for a fraction F of what will be applied later, but it is important to maintain arithmetic progression (rate 1Fa, rate 2Fa, rate 3Fa, etc. where F < 1). As the aim is to determine deficiency thresholds (no application, rate "0") and the optimum content ranges, the first non-zero rate (rate a) needs to be close to the targeted rate to reach maximum yield. When no, or very little, improvement in productivity is found when that rate is doubled (rate 2a), it is a very convincing result for justifying rational fertilization (box 7).

The planting material itself may also be one of the factors studied when different origins have been planted. Trials can be set up for each origin when the area occupied by it is considered large enough to justify greater precision. When the specific needs of different planting materials are investigated, an experimental design with a genetic factor is needed (mineral nutrition × planting material study). Such experimental designs are rare and are mostly of interest for the production of seeds and O × G hybrid materials, for which certain crosses display specific behaviours.

Other types of experimental design exist for other uses, such as "mixture designs" to test products and formulations comprising several nutrients, or "central composite designs", to estimate response surfaces. Such designs can be highly economical in terms of cultivated area (some combinations are not replicated), but their analysis can be complex, and they are sensitive to missing data in some plots and to appreciation errors as regards the negligible nature of certain interactions. This makes them poorly adapted to lengthy trials that cannot be repeated in the event of failure.

Setting up a uniform trial

It takes several years of monitoring to obtain responses in a trial; the responses often depend on the soil type and sometimes on the planting material used. Thought therefore has to be given to priorities for the plantation: What is the most represented soil type?

What material will be planted in the future? Trials are usually conducted at the same time as planting programmes; for efficient fertilization guidance, they need to provide precise information as quickly as possible.

The reason for a fertilization trial is often to determine optimum contents for the most productive period of the plantation, i.e., between 8 and 15 years old. However, trials are usually no longer monitored after 10 years. There is therefore every interest in starting treatments as soon as possible to prevent the reserves present in the biomass of the palms and in the soil from delaying the appearance of a deficiency. A trial usually begins at 3 years old but for some nutrients, such as K and Cl, it is possible to start right from the year of planting. The rates applied in the early years of the trial increase at the same time as the biomass, but the ultimate rates need to be achieved rapidly. A simple protocol always simplifies the synthesis of the end-of-trial results.

Box 7. A very efficient factorial design for oil palm: 3 factors at 3 levels

The frequently large number of combinations to be tested in a factorial design, and the minimum size needed for the unit plots, mean that a perfectly uniform trial area cannot be guaranteed, which can distort the results. Preference should therefore be given to trial designs enabling checks for variations in soil fertility, by setting up incomplete blocks combined with high-rank interactions that it is known can be overlooked. Such is the case with the factorial design combining 3 factors at 3 levels (3^3) allowing a distribution of the 27 combinations (figure 25) in 3 blocks confounded with the interaction of the three factors that rarely has any identifiable agronomic effects. This trial design can be planted without replicating the 27 combinations, since the residual error can be satisfactorily estimated because the interactions between two factors are usually simply linear. This design is efficient due to the quality of the responses observed compared to the size of the area occupied (Yates, 1964 p. 42 and 53).

N0 - P1 - K0	N0 - P0 - K1	N2 - P1 - K2	N2 - P2 - K0	N0 - P0 - K0	N2 - P1 - K1	N0 - P0 - K2	N1 - P0 - K0	N2 - P1 - K0
3	6	9	12	15	18	21	24	27
N1 - P1 - K1	N1 - P2 - K0	N0 - P2 - K2	N1 - P0 - K1	N1 - P2 - K2	N1 - P1 - K0	N0 - P1 - K1	N2 - P2 - K2	N0 - P2 - K0
2	5	8	11	14	17	20	23	26
N1 - P0 - K2	N2 - P0 - K0	N2 - P2 - K1	N2 - P0 - K2	N0 - P2 - K1	N0 - P1 - K2	N1 - P1 - K2	N1 - P2 - K1	N2 - P0 - K1
1	4	7	10	13	16	19	22	25

☐ Block 1　☐ Block 2　▨ Block 3

Figure 25. Factorial design with 3 factors at 3 levels comprising 27 unit plots

The number of each experimental plot is in the bottom right-hand corner and the combination of the 3 factors is shown in the middle of each box. Plots 1 to 9, 10 to 18 and 19 to 27 have been grouped into three incomplete blocks in which the treatments have been chosen to enable a satisfactory analysis of the results. There are 3 combinations in each block with N0, 3 combinations with N1 and 3 combinations with N2. The same applies for the P and K factors. During the analysis of variance, this particularity enables calculation of a block effect that will reduce residual variance accordingly.

The experimental plots of a trial are compact groups of palms to which different fertilizations are applied, such as the 27 combinations in a factorial trial (3^3). The basic observations (yield, leaf analyses, etc.) come from the useful palms located in the middle of the experimental plots and surrounded by border palms receiving the same fertilization (figure 26). The size of the experimental plots is key to the trial design. Firstly, for practical reasons (work volume, risk of error, difficulty in ensuring uniform soil conditions over a large area), the aim is to set up trials on small areas. Secondly, it takes a minimum number of palms for satisfactory observation of how a combination of treatments affects yields. For instance, a set of 9 useful palms per experimental plot is a good compromise in a standard situation. That number may be larger because some individual palms may disappear before the end of the trial; the area of a trial varies depending on the health context. For example, a factorial trial with 3 factors on 3 levels (3^3) with plots of 5 rows of 5 palms (of which 9 useful palms) occupies 4.7 ha (143 palms/ha). A (3^3) factorial trial with plots of 6 rows of 7 palms (20 experimental palms) occupies 7.9 ha.

Trial uniformity is crucial for obtaining quality results and precise responses. Choices are made at each stage of trial preparation to guarantee such uniformity:
– production and culling of seedlings at the end of the nursery stage so as to plant only "normal" and "homogeneous" palms,

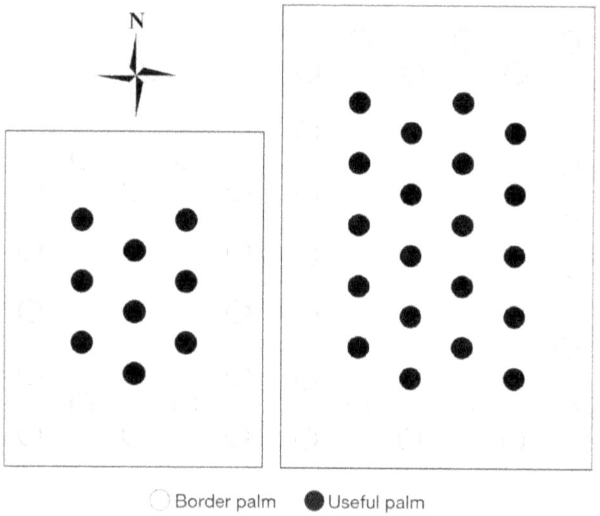

Figure 26. Two examples of different-sized experimental plots. Plot 1: 25 palms; plot 2: 42 palms

All the palms in the experimental plots are fertilized according to the protocol. Border palms limit the effects of the treatments from one plot to another, on the useful palms used for observations. Plot size depends on the health context, the size of the available uniform area for the trial, but also on other criteria, such as the organization of individual weighing for harvested bunches. All the palms must be labelled for error-free identification, plot boundaries are marked out at each corner with signs.

– choice of plot: a single soil type and, if possible, a single type of previous plant cover on a flat, or failing that, gently sloping plot,
– pest control, to avoid seedling losses and retarded growth in the early years. There are foreseeable border effects to be avoided, such oil palm root miners (*Sagalassa valida*) living in forest borders, or rodents on the edges of bottomlands.

If the trial is set up in an existing crop a vegetative growth indicator, such as frond length, can help in detecting fertility gradients existing at the outset of the trial and in finding the best layout for the trial (figure 27).

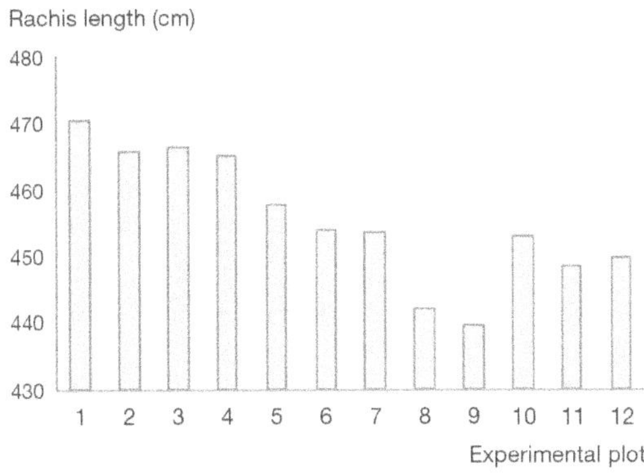

Figure 27. Average length of rank 17 frond at 5 years, per experimental plot, before the start of the protocol

> The trial is intended to test the application of boron at two levels (presence-absence) with 6 replications. The experimental plots are located along the main axis of the block corresponding to a vegetative development gradient. The replications therefore need to be located along that axis.

Despite these precautions, variations in soil properties are often difficult to detect, so it is important to favour uniformity inside the blocks (complete or incomplete). If a gradient is suspected (slope, main axis of the plot), the blocks are laid out along that gradient. The (3^3) factorial design can be divided into three blocks, so that this precaution can be respected (figure 28).

When setting up a trial, palm labelling and identification of the treatments to be applied in each experimental plot must not be overlooked. Precise collection and recording of data should be organized to ensure their conservation over the full lifespan of the trial.

Aggregating data and determining local optimum contents

For each combination of factors studied, or each treatment applied in each experimental plot, the useful palms are used to determine the contents of a composite leaf sample by analysis, so as to calculate average yields over a given period based on individual bunch weights, and to calculate the mean values of vegetative observations (frond emission and length). Even though precautions are taken when strictly culling

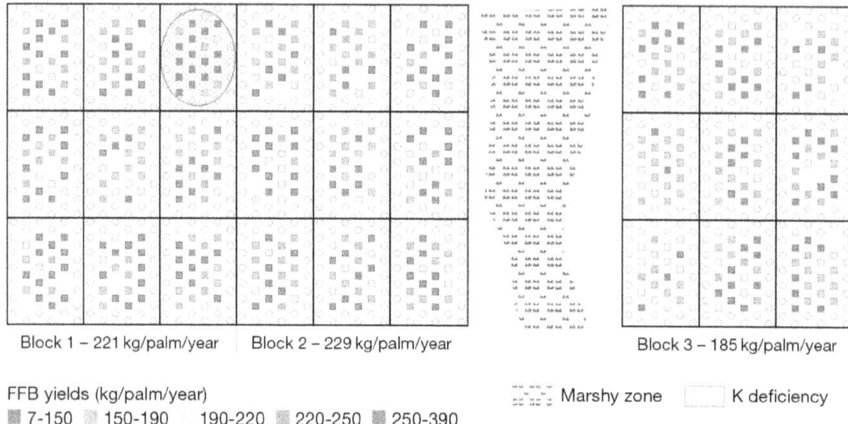

Figure 28. Example of a (3^3) factorial design in which the experimental plots have been grouped in 3 blocks

The average yields from 9 to 11 years revealed a significant block effect reflected in an average yield in block 3 (185 kg/palm) below that of the other two blocks (block 1, 221 kg/palm; block 2, 229 kg/palm): the lowest individual productivity category is shown in red. The plot was flat and a hydromorphic zone was avoided between blocks 2 and 3, but despite that precaution variability between the blocks could not be detected before the trial was set up. Another particularity also appeared when observing leaf K contents: in an experimental plot of block 1 the drop in K contents and the lower yields of the palms were explained by the lower soil K reserves than in the rest of the trial.

seedlings in the nursery, off-standard palms may appear after a few years, usually at the start of bearing. It mostly involves abnormalities (stunting, abortive palms) or diseases, and those palms are not included when calculating the average effects of the variables observed in each experimental plot. Nonetheless, they need to be identified respecting strict standards. Deficiencies occurring in trials have an impact on the vegetative appearance of the palms, which must not be confused with a genetic abnormality. Statistical analyses are usually carried out on annual plot values (composite sample contents, or average yields per palm). The means obtained for the treatment of each factor are used to keep track of how mineral nutrition evolves over time and to understand how deficiencies that limit yields become established (figure 29).

Experimental precision

For each fertilization trial the minimum detectable difference with a reasonable probability (MDD) can be calculated retrospectively. That probability (also called power) is usually fixed at 75%. The calculation is of interest when there is no significant effect of the treatments on yields even though the treatment has been applied for several years. Rather than concluding that the factors do not have any effect on yields, it can be concluded that the differences between two treatments are probably under the MDD. If the latter is low, recommending one treatment rather than another will have little impact.

In the opposite case, it is wise to test the factors again in a more appropriate experimental design (uniform plantation plot, larger number of palms, low mortality, etc.).

Given the substantial variation in MDD from one trial to another, it is best to express it as a percentage of the average trial yield. For example, the MDD was calculated for a power of 75% over 15 lengthy trials conducted in different countries for up to 21 years for the longest. It varied from 7 to 30 kg of bunches per palm per year because potential yields varied considerably depending on the agroecological situations. This investigation also showed the MDD to be rarely below 5% of the yield (1 out of 15 trials), whilst MDD values of between 5 and 10% of the yield were often found (9 out of 15 trials), which seemed to be the norm for considering that a trial had been satisfactorily conducted. On the other hand, the results from 5 trials in our survey with MDD values over 10% did not provide any precise information on how the factors being studied impacted yields. It would be a good idea to set up some new trials under more controlled conditions.

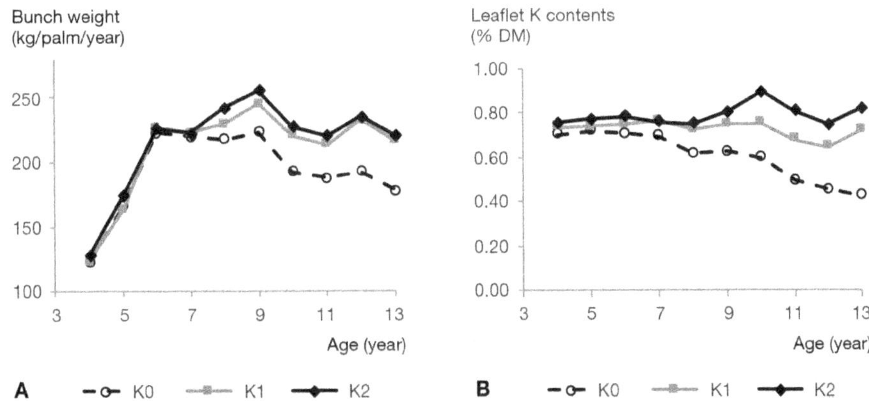

Starting from 6 years, the KCl application rates were: K0, control; K1, 1 kg/palm/year and K2, 3 kg/palm/year.

Figure 29. Trends for annual bunch production (figure 29A) and leaf K contents (figure 29B) in the factorial trial with 3 fertilization factors (N P K) at 3 levels in Ecuador

Starting from 8 years, yields in the K0 experimental plots began to fall compared to those in K1 and K2 and the differences became significant from 9 years onwards. K0 always displayed the lowest K contents; from 8 years onwards, the differences when compared to the K1 and K2 became significant.

Drawing up the fertilizer schedule from the experimental results

This stage is fundamental, as the aim is to transform statistical results (usually an ANOVA) into a fertilization decision. It is first of all necessary, nutrient by nutrient, to pinpoint the optimum content ranges based on the trial results. These optimum contents are those beyond which yield gains due to the fertilizer are judged too low to be of interest. Next, the fertilizer schedules making it possible to achieve those objectives in each fertilization sector will be drawn up.

Determining the optimum content range per nutrient

Two approaches are possible depending on how the trial results are processed.

"Observation of tipping points" method

This method is applied by seeking a period in the trial when significant differences in yields are found between treatments, and the leaf contents obtained with those treatments are examined. Particular attention is paid to the following changes:
– when the treatment studied, e.g., K0, does not result in yields that can be achieved with another treatment (K1 or K2) and the difference becomes significant, the K content associated with K0 typifies a deficient status. The optimum content range for potassium must therefore be higher than the deficient content,
– when the difference in yields between two fertilized treatments, e.g., between K1 and K2, is not significant it is considered that a satisfactory nutritional status is virtually achieved with the first treatment and can serve as a reference.

Observation of these changes derives from the dynamics of the trial results. It is identified from what content level a loss in yield becomes statistically measurable. The variation in leaf content and its effect on yields is staggered by at least a year, and usually by two years.

We illustrate this approach by taking another look at the example in figure 29. The drop in leaf K content from 0.70 to 0.60% DM between 7 and 8 years would seem to be the cause of the difference in yields between K1 and K0, which became significant from 9 years onwards: the K deficiency threshold would therefore seem to be between those two values. The optimum content range can therefore be set at a minimum of 0.65% DM – or even 0.70% DM as a precaution. It was also found in this trial that for K1 (1 kg KCl/palm/year) leaf contents were stable over the duration of the trial, with a mean value of 0.73% DM. There was therefore no point in exceeding that content as the yields recorded for K1 and K2 were not different. That content therefore fell within the optimum content zone, which could therefore be set at 0.70-0.75% DM, and used to draw up the fertilizer schedule.

Trials also indicate fertilizer application rates suited to achieving our objective. In our example (figure 29), the K1 rate of 1 kg KCl/palm/year could be proposed to maintain leaf contents at between 0.70 and 0.75% DM. Rate K2 (3 kg KCl/palm/year) should only be proposed to correct very low contents: it is not applicable over long periods for economic reasons.

The method proposed depends on the possibility of observing one or more significant differences in yields between treatments, and the quality of the trial data is paramount for translating that into leaf content thresholds.

Experimental results are not always significant and some other decision-making rules may be used, even though they remain subjective. An increase in yields between two treatments under 5%, such as between K1 and K2, is not usually significant. Conversely, if a treatment leads to an increase in yields of 10% or more that gain is most often significant; if that is not the case, the quality of the trial data needs to be checked.

Foster (2003) referred to a variation in yields in tonnes of bunches/ha/year: below a variation of 0.5 tonnes/ha/year, there is no point in modifying a treatment for another; for a variation of between 0.5 and 2 tonnes/ha/year an adjustment should be sought; beyond a drop of 2 tonnes/ha/year it is essential to correct the deficiency that is limiting yields.

Whatever the method adopted to convert experimental results into a fertilizer schedule, it is necessary to examine the precision of the results and calculate the MDD to draw satisfactory conclusions (see previous section "Experimental precision"). In fact, if there are no significant effects between treatments after several years of observations it would be risky to assume that the tested factors have no effect on yields. Differences in yields between treatments can be high without necessarily being significant, due to results that lack precision.

"Modelling of experimental results" method

The optimum content range can also be determined by mathematical transformation of the mean trial results. This is done by modelling the responses of yields and leaf contents to fertilizer application rates and precisely calculating the contents that correspond to an economic return threshold that has been fixed for the fertilizer.

The first stage consists in identifying a period when the effects are considered to be expressed (figure 29: 10-13 year period). This period represents a satisfactory average yield achievable throughout the plantation in a mature crop from 10 years onwards.

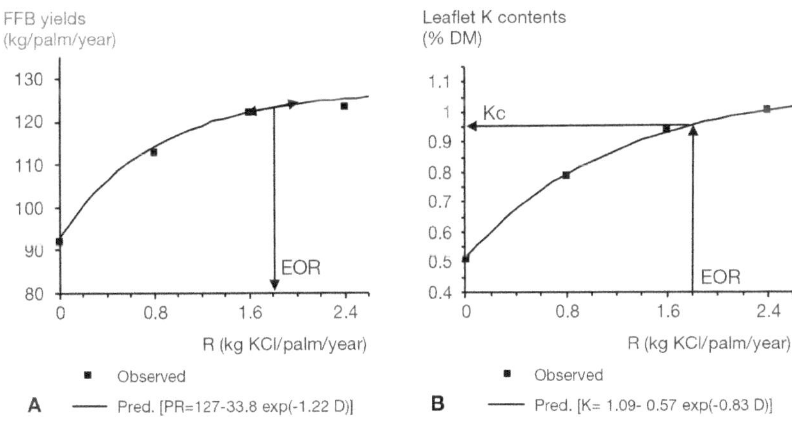

Figure 30. Example of curves for yield (Figure 30A) and leaf content (Figure 30B) responses to fertilizer rates (D) (according to Caliman *et al.*, 1994)

The equations were fitted to the Mitscherlich model. At every point along curve A, the value of the tangent indicates the kg bunches:kg fertilizer ratio. It decreases with the KCl applications up to a value that remains economically acceptable. The corresponding economically optimum fertilizer application rate (EOR) is determined; it is 1.7 kg KCl/palm/year. Figure 30B indicates the leaf potassium content obtained with this fertilizer rate. The value Kc = 0.95% DM will serve as a reference to manage potassium nutrition in the plantation where the trial was conducted.

The second stage consists in calculating the equations of the leaf content and yield response curves for the trial fertilizer rates by adopting, for example, a Mitscherlich model:

$$Y = a - b \, \exp(cX)$$

Y is the dependent variable (yield, leaf content); X is the independent variable (fertilizer rate); a, b and c are constants.

An economically optimum fertilizer application rate (EOR) is determined from the yield response curve taking into account the average economic parameters. This optimum balance approach was described by Caliman *et al.* (1994, figure 30). The target leaf content, or "critical level" is also determined, along with the fertilizer rate to achieve that content. The fertilizer schedule is drawn up from these values.

This approach can be generalized to response surfaces by combining the factors two-by-two. Mathematical processing of the response surfaces can be used to calculate the application rates for each nutrient and the corresponding leaf contents depending on the economic parameters adopted by the user. A full example of trial processing was provided by Webb (2009).

These calculation methods are based on strict criteria, but they require a realistic projection of input costs and of the palm oil market over long periods, because it takes a year or two for the oil palm to respond to fertilizers. The simulation carried out by Webb (2009) shows that the oil price and cost of inputs greatly influences the expected yield and the fertilizer application rates required to achieve it. Proper parameterization is therefore needed.

Drawing up a fertilizer schedule from the optimum content range

Once an optimum value has been determined for a given nutrient, a range of optimum contents will be defined around that value and a fertilizer schedule will be drawn up. It comprises content categories that advocate specific fertilizer application rates. The most important category is the one containing the optimum content Kc = 0.95% DM in the case of figure 30. To keep potassium contents within that category, the economically optimum fertilizer application rate (EOR = 1.7 kg KCl) will be applied. If the leaf content of the sample is located in a lower category, a fertilizer rate above the EOR will be recommended, and vice versa (table 5).

Table 5. KCl fertilizer schedule (kg/palm/year) depending on the leaf sample content

K contents (% DM)	< 0.8	0.8 to 0.9	0.9 to 1.0	1.0 to 1.1	> 1.1
KCl rate (kg)	2.7	2.2	1.7	1	0

The fertilizer schedule is centred on the range of optimum contents and the economically optimum fertilizer application rate determined from the trial results. When contents are in high categories the fertilization recommendation is lowered, even up to no longer applying any fertilizer for a few years if necessary. On the other hand, the lowest content ranges recommend higher application rates. It was chosen here to cap the maximum rate at 2.7 kg KCl/palm/year.

However, Webb (2009) specified that using a fertilizer schedule is only suitable for plots where the planting material and soil conditions are similar to those of the trial. This comment is paramount, because it has consequences for experimental data requirements and for the construction of fertilizer schedules:

– if there are several soil units with physical and chemical properties that are different enough to influence mineral nutrition, trials need to be set up on the largest units. This precaution is important because, on some soils, the response of leaf K contents to KCl applications can be distorted by the abundance of calcium in the soil, as we shall see later,

– when the experimental network is young, with few reference contents to guide fertilization, there is no other choice but to extrapolate results to the entire plantation. So, the leaf contents of the plantation need to be analysed spatially to detect those whose response to fertilizers does not comply with the experimental results. It is thus possible to moderate fertilizer applications if a problem of uptake or translocation of nutrients to the leaflets is suspected (box 8).

Box 8. Cost-effectiveness of fertilizer applications and environmental protection

It is not uncommon to find that the yields recorded in a trial continue to increase when switching from one treatment to another, even if the difference is not significant. There may therefore be the temptation to achieve the highest leaf contents in the trial and apply the highest fertilizer application rates, but it is best not to target the maximum yield obtained experimentally, but to bear in mind what becomes of the fertilizer for the highest rates.

If mineral nutrients are only slightly taken up by the crop, or not at all, what happens to them? If they accumulate in the vegetative organs of the plant, they return to the soil via pruning or at the time of replanting when old stems are felled. It is to be feared that sooner or later they will be in excess compared to the soil storage capacity. So, as a precaution for the environment, it needs to be planned in the fertilizer schedule that fertilizer applications will no longer be needed beyond a given leaf content.

Applying the conclusions of the experimental approach

Fertilization trials rapidly detect the key mineral nutrients that limit yields in a given context. They indicate deficiency thresholds, which will be used to determine the minimum leaf contents to be achieved in all the leaf sampling units. They also specify from what point fertilizer application rates increase leaf contents without improving yields.

Further modelling of the experimental results is possible using response curves or surfaces, which are used to obtain precise norms, valid in trial plot contexts; it is considered that theses curves/surfaces can be extrapolated to fertilize leaf sampling units (LSU). However, the observed responses do not only depend on the fertilizer application rates studied, but also on the properties of the soil and its ability to store and give back mineral nutrients. Consequently, spatial studying of content responses inside soil units is a tool that is just as important as the mathematical processing of trial results. For the LSUs in a plantation, it involves checking that the observed leaf content responses to fertilizer applications comply with the responses obtained experimentally.

A better understanding of trial results

A trial is not only used to determine optimum leaf contents. It is a uniform reference site where cultural practices are effectively controlled (upkeep, fertilization, harvesting). A trial sometimes provides answers to questions that arise from the plantation, where data are not organized in such a way as to express explanatory hypotheses. Let us take a few examples.

Over a large area of the plantation unexpected variation can be seen in N, K or Mg contents to an unusual extent. A trial is clearly the best way of checking whether that variation concerns all the application rates tested when the nutrient in question is one of the factors being studied. A few examples have shown that such events are often independent of fertilization. Consequently, the fertilizer schedule will not be applied in its entirety if the experimental results confirm that the variations in leaf content are due to factors other than fertilization.

Likewise, a trial is a good control when a drop in yields is observed either in a given year, or several years running. The role of the nutrients being tested by the protocol can be examined. This thus provides some very precise information about yield components (bunch number and average bunch weight). However, the hypotheses put forward for the mechanisms underlying the drop in yields will not be the same: a drop in bunch number usually occurs after stress, whilst a drop in average bunch weight can be due to a pollination problem (figure 3).

Trials also provide information on the resilience of soil reserves and the risk of a deficiency occurring. Often, in the early years of a trial no positive or negative effects of the factors being studied are detected on yields (especially fertilizer applications).

In the example shown in figure 29, the potassium nutrition of the K0 treatment (without KCl fertilizer) slowly deteriorated; it was only from 9 years onwards that significant differences in yields were recorded compared to K1 and K2. Depending on the soil type the time taken for deficiencies to occur is often between 5 and 10 years for controls without N or K fertilization (Dubos and Flori, 2014). Consequently, in a regularly fertilized commercial plantation the risk of inducing a loss of yields following an underestimated fertilizer recommendation is very low, as the decline in contents is seen before it affects yields. Trials also provide a good estimation of the reserves available for older crops when their fertilization is halted two to three years before replanting.

4

Extrapolating fertilizer schedules resulting from trials

This involves determining the geographical areas over which the results of the fertilizer schedule are applicable. Fertilization trials provide response curves for how yields and leaf contents react to fertilizer applications. After identifying deficiency thresholds and optimum contents, fertilizer schedules adapted to local soil and climatic conditions are drafted.

It needs to be checked that the optimum contents and the fertilizer schedules obtained are valid for the entire plantation. To that end, characterizing the soils of the plantation and a spatial analysis of plot data are powerful ways of detecting zones in the plantation that are an exception. The fertilizer schedule will have to be adjusted for those sectors, and that may mean setting up new trials.

Analysing the plantation reaction on an LSU scale

Planting material origin and soil type are the main factors that can affect nutrient allocation in leaflets, as climate data (rainfall and sunshine) do not vary enough to affect mineral nutrition inside a plantation.

As things stand, variability in leaf contents depending on the planting material is poorly documented. However, it has been established that for equivalent fertilization, the N and K leaf contents of *E. oleifera* × *E. guineensis* palms are lower than those of *E. guineensis* palms at the same site, due to different foliage and stem biomass. As soon as the planted areas become large (e.g., over 1,000 ha) trials need to be set up for each planting material type/genetic origin. The same applies for crosses that have at least one parent with a genetic origin recognized as being different from the rest of the planting material (e.g., between La Mé and Yangambi for *E. guineensis*, and between Coari and Manicoré for *E. oleifera*).

The chemical properties of soils can modify leaf contents through mechanisms that are not yet well understood, but whose consequences can be important, especially for potassium ("Taking into account soil calcium contents when using KCl" page 61). As soon as data are available (soil map, topography map) showing that large soil units (e.g., alluvial lowlands and sedimentary rock plateaux) have different texture and chemical properties, it is essential to design specific fertilization trials.

What information can be drawn from soil analyses?

A detailed soil map is not always available, but it is common to analyse a few composite samples before planting the first palms. The samples are usually representative of the main soil units and provide two types of useful information:

– the nutrients (N, P, K and Mg for the most part) for which soil reserves are low and for which deficiencies may occur in the early cropping years. The first fertilizer schedules will take that situation into account,

– the properties affecting the ability of the soil to store and give back water and mineral nutrients; this mainly concerns the organic matter (OM) content, the cation exchange capacity (CEC) and the texture. If these properties are sub-optimal, an attempt will be made to improve soil functioning by appropriate cultural practices, such as planting a legume cover crop and providing organic matter (see "Improving the physico-chemical properties of soils", page 73).

Soil analyses prior to planting will help detect a likely phosphorus deficiency. The reference thresholds below which a response to phosphate fertilizers is obtained vary depending on the analysis methods (according to Caliman *et al.*, 1994):

– total P, 400 ppm
– Saunders P, 130 ppm
– Olsen P, 30 ppm
– Bray-2 P, 15 ppm.

When analysis results are below these thresholds, phosphates can be systematically applied until the experimental results define the optimum leaf contents and the recommended application rates. Sometimes, a share of the P provided is fixed in a form with low solubility, especially when the soil pH is acidic, and in the presence of aluminium and iron oxides. It is first necessary to saturate the fixing capacity of the soil before content and yield responses to fertilizer applications are obtained. Fertilization trials are the only way of carrying out such observations and specifying the cost-effectiveness of P fertilization under such conditions.

Soil exchangeable cation contents (mainly K^+ and Mg^{++}) do not always provide reliable indications for detecting a deficiency and, here again, the analysis method is important. Exchangeable K and Mg reserves are considered to be insufficient when under 0.2 cmol/kg, but that threshold refers to extraction with ammonium acetate, which is inappropriate for acid tropical soils. For instance, Foster and Prabowo (1996) showed that K extraction with boiling HCl gave an estimation of K reserves coherent with the yield responses observed in a network of trials in Indonesia, while extraction with ammonium acetate did not explain the results. The recommendation is therefore to use a method for which the pH of the extraction solution (cobalt hexamine trichloride or barium chloride) is similar to the pH of the soil. Even taking this precaution, exchangeable K analysis does not always provide a satisfactory estimation of what is available for oil palms. For instance, on alluvial soils in Colombia, Dubos *et al.* (2011) concluded that the K reserves ought to have been much higher than indicated by extraction with cobalt hexamine trichloride (contents under 0.2 cmol/kg in the absence of K application). This conclusion was established by way of a potassium balance of the aboveground organs (table 6).

Table 6. K and Cl weights (kg/palm) measured in 11-year-old *guineensis* type palms in Colombia

Palm organ	Cl stock (kg)			K stock (kg)		
	T0	TKCl	TNaCl	T0	TKCl	TNaCl
Stem	0.48	1.76	1.91	3.27	4.10	4.51
Leaf crown	0.39	1.23	1.22	1.37	1.46	1.44
Total	0.87	2.99	3.13	4.64	5.56	5.96

The quantities were calculated from a stem biomass of 150 kg of DM and from 35 fronds in the crown. Chlorine was provided in the form of KCl (TKCl) or NaCl (TNaCl) at a rate of 2 kg Cl/palm/year. Compared to the control without fertilizer (T0), the Cl stock more than trebled with TKCl and TNaCl, and that was accompanied by an increase in K stock. For TNaCl, the K stock increased by 28% compared to the control and that K could only have come from the soil reserves, which the soil analysis indicated to be low.

All in all, soil analyses only provide limited information about fertilizer requirements. As they are not well correlated with the applications of fertilizing elements and with yields they cannot serve directly as a fertilization guidance tool. However, they do provide some essential information for understanding leaf content responses and how the leaf analysis tool functions.

It has been seen that calcium is not an important nutrient for oil palm, but the abundance of calcium in the soil can interfere with potassium allocation in leaflets and distort the contents analysed. However, potassium fertilization accounts for the largest share of fertilizer applications, which warrants taking a separate look at this question specific to KCl applications.

Taking into account soil calcium contents when using KCl

Potassium is the major fertilizing element for oil palms, and the correlations obtained between yields and leaf contents have greatly contributed to the success of leaf analysis. Potassium is also the nutrient for which it has been difficult in certain circumstances to specify the quantities of fertilizer needed. Those difficulties provide a good example of the interaction between soil properties and nutrient allocation in leaflets.

Dubos *et al.* (2017a) studied how leaf K contents responded to KCl applications using the results from 13 fertilization trials spread throughout the world and located on different soil types. They classed the responses observed in three categories (figure 31): the K content increased in line with the fertilizer application (positive response, which is the case of most trials), the content did not vary (neutral response), and the content fell with the application (negative response).

When leaf K contents do not vary or fall, it is impossible to determine an optimum K content by the response curve method described in the "Determining the optimum content range per nutrient" section, page 54. The soil property best correlated to this dysfunctioning is "calcium pressure", expressed by the ratio (as a %) between the

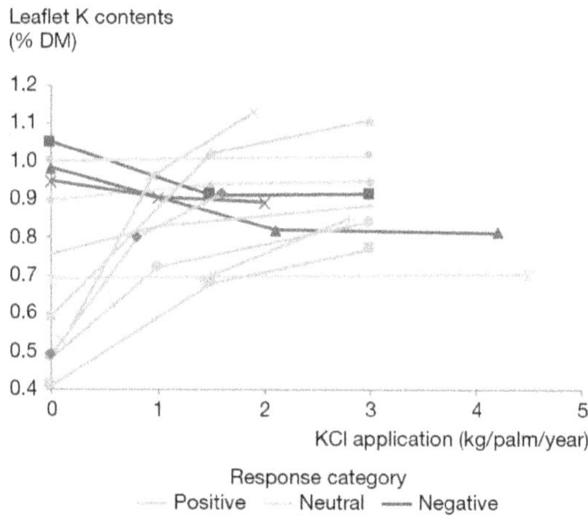

Figure 31. Three categories of leaflet K content responses to KCl application (trials located in Africa, Asia and Latin America)

> For 7 out of 13 trials, the contents rose in line with the application (green lines); for 3 trials, there was no response (yellow lines); for 3 trials the contents fell (red lines). The data from these trials were used to draw up figure 32.

soil exchangeable Ca content and the cation exchange capacity (CEC). This ratio is used to predict the increase in leaf K contents compared to a control without KCl fertilization (figure 32).

In two of the trials used to establish these relations, Dubos *et al.* (2011) effectively showed that leaf K contents decreased with KCl rates, which did not cast doubt on K uptake by the oil palms because the stem and leaf rachis contents increased (figure 33).

When soil calcium pressure is high, it is found that at the same time as the drop in leaf K contents with KCl applications, there is also an increase in Ca contents. This phenomenon, known as K-Ca antagonism, also brings into play Cl-Ca synergy, which is found when applying NaCl. It is not an antagonism involving preferential Ca uptake by the palm over K, but rather for preferential Ca allocation over K in leaflets.

The determinants of cation allocation were not precisely described, but it turns out that when calcium pressure is high, leaf K contents can stagnate within low leaf content value ranges (0.70-0.80% DM), without really being able to talk about a deficiency, because it does not reduce yields. However, these soil conditions prevent a satisfactory recommendation from being made for KCl fertilizer applications to raise leaflet K contents.

Detecting distortions due to soil properties

In the previous example, the safest method was to rely on soil analyses, which revealed sectors in the plantation where the Ca:CEC exceeded 50%. However, when a soil map is available, the description of the soil units is only based on a limited number

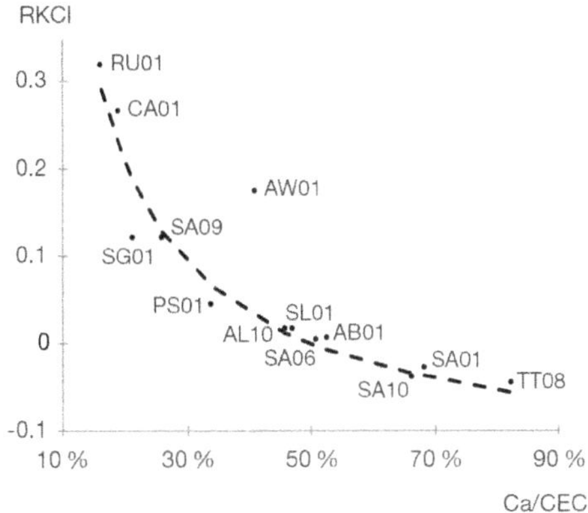

Figure 32. K content response curve depending on calcium pressure
(according to Dubos *et al.* 2017a)

The data in figure 31 were used to calculate the response R KCl which expresses the variation in K content (% DM) per kg of KCl applied, between the highest KCl treatment and the treatment without KCl. When "calcium pressure", i.e., the Ca:CEC ratio (soil exchangeable Ca content divided by the cation exchange capacity) reaches then exceeds 50%, the responses of the leaf K contents to KCl application become neutral (R KCl = 0), then negative.

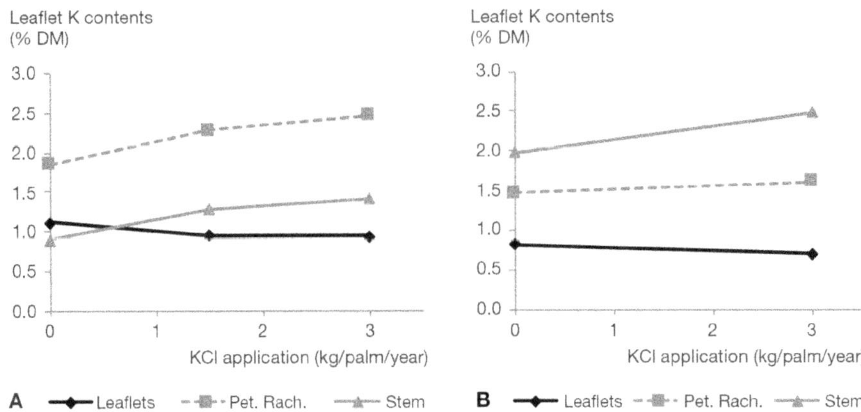

Figure 33. Potassium contents found in trials in Ecuador (figure 33A)
and in Colombia (figure 33B), at 10 and 11 years old

In these two plantations, the analyses indicated that KCl applications significantly reduce leaflet K contents. However, the increase in contents for the stems and the petiole + rachis combination (Pet. Rach.) of frond 17 showed that there was effective potassium uptake by the roots and migration to the leaf rachises.

of analyses. There is then a need to examine the Ca:CEC ratio of the soil in each LSU of the site from which leaflet samples will be taken for leaf analysis.

Once the sectors where the Ca:CEC ratio is sufficiently high to distort K allocation in the leafets have been identified, it is then necessary to confirm that leaf contents do not respond to KCl applications. An analysis of the performance of the LSUs with mapping tools helps to confirm the existence of soil conditions that alter potassium allocation, but without any consequences for yields. These tools can be applied to any nutrient where the aim is to establish that fertilizer recommendations really do have an effect on leaf contents and to specify the extent of that effect (see "Constructing a geographic information system", page 65).

Lastly, simple response tests can be set up to confirm that problems exist in the interpretation of leaf analysis results localized in certain zones. Such tests are both less expensive and faster than setting up a true fertilization trial (see "Setting up reactivity tests", page 66).

Associating a reference soil analysis with each leaf sample

The area covered by the leaf sample should be restricted to a part of the leaf sampling unit (LSU) occupied by a single soil unit (see "Choosing the palms of the reference sample inside the LSU", page 41). It is therefore perfectly legitimate to have a reference analysis for that soil. It is thus possible to understand how contents respond to fertilizers when dysfunctioning is suspected, as already seen. This will then provide sufficient datasets (leaf contents, fertilizer applications and soil properties) to consider a multivariate analysis and identify some explanatory relations to specify optimum contents depending on the soil.

The reference soil analysis should be considered as a descriptive dataset for each LSU. It is not worth repeating it over time (a sample per cycle) as it is not intended for monitoring soil reserves (subject covered in the "Assessing soil reserves" section, page 70). Neither should it be subject to variations in contents due to fertilization practices: samples should therefore be taken outside mineral or organic fertilization zones. When the crop has already been in place for at least ten years, zones receiving pruned fronds should be avoided as the soil may be enriched in cations through biomass recycling. All in all, it is preferable to analyse composite soil samples as soon as possible, i.e., at 3 years when establishing the leaf sample for the LSU. Highly localized soil sampling, of which the best example is the paired rows design (figure 23, LSU02) is recommended to obtain a reference that will be associated with a one-off object for which geographical coordinates can be specified: each palm in the sample can be associated with an elementary soil sample to make up the composite. A depth of 30 centimetres is enough to characterize the superficial layer where the surface roots located more than 2 metres from the stem are concentrated. The underlying horizons can also be sampled over the same thickness down to 90 centimetres if more information is required on the plantation soils.

Constructing a geographic information system (GIS)

For a finer understanding of the relations existing between soil properties and leaf contents, independently of the fertilizer rates applied, numerous soils would have to be tested experimentally in the plantation for several nutrients (P, K and Mg as a priority). Such a method is lengthy and costly. On the other hand, by examining the responses of leaf contents in the LSU areas, it can be detected where the expected content is not achieved by the fertilizer schedule. If a GIS is available, each leaf analysis observation point can be entered and an attempt will be made to understand if the lowest values and the highest values are randomly distributed or geographically grouped. It may be necessary to take age into account as presented in figure 34, which describes the average leaf contents for magnesium between 8 and 12 years old. The chemical properties of the LSU soils can also be mapped to determine whether the sectors where the Ca:CEC ratio is higher are superimposed on the lowest K contents at the same age.

Spatial analysis can be used to superimpose several layers of information (geological map, soil map, topographical map, soil water reserves map, water table depth, planting material). Such an approach is very useful for putting forward hypotheses, not only on what is disrupting leaf contents for whatever nutrient, but also on the role of other variables (e.g., health status), to explain yield variations.

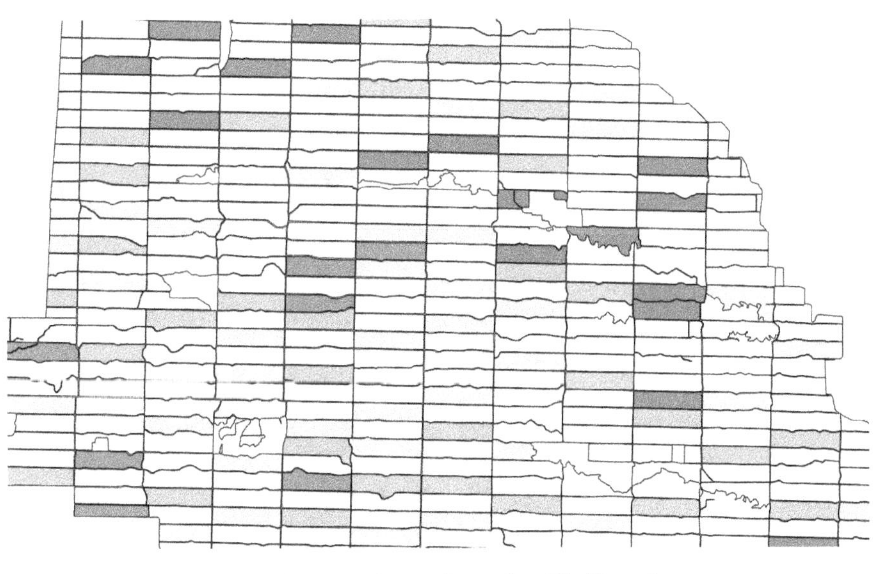

Figure 34. Mapping of average Mg contents between 8 and 12 years

The contents of each LSU were entered in the plot from which the sample was taken. A few soil samples confirmed that high Mg contents (South-West of the plantation) coincided with chemical properties that differed from those in the other soil units. In places where the contents were lowest, uptake tests confirmed that Mg translocation to the leaflets was blocked.

Setting up reactivity tests

The leaf rachis is a reference material for testing potassium uptake, as its contents increase with KCl applications, even when leaflet contents decrease with applications (example, figure 33). These contradictory responses enable reactivity tests to be set

Box 9. Reactivity tests for rapid checking of leaf content sensitivity in different zones of the plantation

The simplest experimental design consists in delineating experimental plots of around 20 useful palms surrounded by border palms and applying two levels of KCl (a control without fertilizer and 3 kg/palm/year), with 3 replications, i.e., 6 experimental plots in all (figure 35). Leaf samples are taken before the first application of the treatments, then two months after KCl application over three years. It is necessary to analyse the leaflets and the frond rachises at the same time and observe the results for each organ. If no improvement in leaf contents is detected for the fertilized treatment, it is likely that K translocation in the leaflet is not properly taking place.

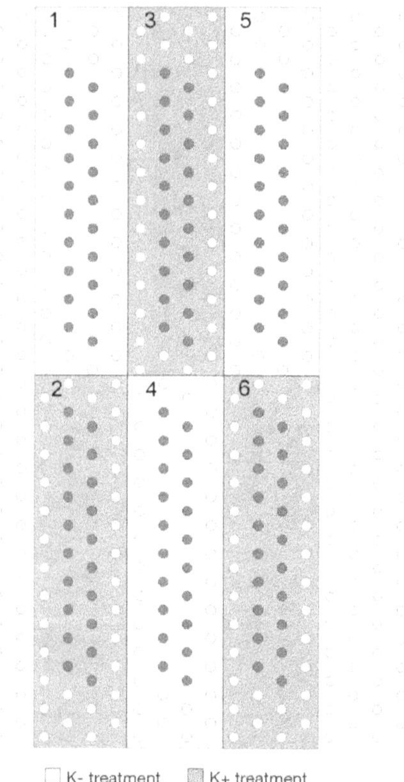

A uniform zone of a plot was isolated to delineate a field of 6 experimental plots comprising 4 rows of 13 palms. Experimental plots 1 and 2, 3 and 4, and 5 and 6, formed 3 replicates. After randomization, all the palms in each experimental plot were allotted one of the two treatments (K^+: 3 kg KCl/year, K^-: without KCl fertilization). Some leaf samples were taken using the useful palms (in the middle) and the leaf contents were monitored over three years.

Figure 35. Example of a reactivity test for leaf K contents

up to confirm that the K contents sought will not be achievable in some LSUs, whatever the K rate applied. These tests are faster and less expensive than resorting to a standard trial (box 9).

Conditions for which a leaf analysis can no longer be used satisfactorily are not common and mostly involve potassium. When they do occur, another solution has to be found to establish a fertilizer recommendation. Foster and Prabowo (1996) proposed using rachis analysis to establish a diagnosis; they considered a K content to be satisfactory (absence of any deficiency having an impact on yields) when over 1% DM. It could therefore be concluded that for the two trials presented in figure 33, the nutritional status at 10 and 11 years old was satisfactory without any KCl application.

Another alternative consists in defining a maintenance application rate, which can be established from chlorine contents when it is necessary to check chlorine nutrition. That rate can also be calculated to compensate for some or all of the potassium exports in bunches, but the estimation calls for precise information on their composition, which is not always available. It also has to be modulated in line with the potential yields of each LSU, which are not always easy to specify. Consequently, for economic and environmental reasons it is recommended that rates exceeding 1.5 kg KCl/palm/year should not be applied.

5

Adopting sustainable fertilization practices: prospects and recommendations

By having access to a decision-support tool for oil palm nutrition that is parameterized according to local conditions, optimum fertilizer rates can be applied to achieve potential yield, while keeping control over the cost of this productivity factor. The process is perfectly effective when the fertilizers applied in a plantation produce the same effects as in trials and enable the recommended leaf contents to be reached. Working conditions in trials are clearly more uniform and controlled than in plantation LSUs, so it is necessary to adopt practices that enable optimum uptake of mineral nutrients at every point in the plantation. If that is not the case, larger fertilizer application rates than originally planned will have to be provided. Economic consequences are not the only challenge to rational fertilization, since the quantities of nutrients not taken up by the crop can generate undesirable effects in terms of soil health and the quality of water resources.

It is therefore only worth investing in a decision-support tool if the best cultural practices are adopted.

Preserving soil health

Initially, oil palm fertilization was especially seen as a way of maximizing yields. Then, with the steady increase in the cost of inputs derived from the mining extraction and chemical industries, economic considerations were introduced, the aim being to achieve high yields at an acceptable fertilization cost. Today, it is essential to target a much more ambitious objective, which consists in reconciling agronomic intensification with ecological optimization. As recalled at the beginning of this guide, soil physico-chemical properties largely govern the potential yield of each plantation. Proceeding in such a way that fertilization does not alter the capacity of soils to ensure ecosystem services enabling satisfactory water and nutrient supplies has to be a long-term endeavour. Thought therefore needs to be given to the consequences of fertilizer recommendations that are not in phase with the requirements of the crop: Could this modify the properties of fertilized soil? Conversely, could there be a gradual depletion of the soil that jeopardises its sustainability?

Caring for the chemical fertility of soils

The aim is to ensure that there is no detrimental change in soil properties caused by an accumulation of a given mineral nutrient in the soil. Such an imbalance can lead to impoverishment of the soil in other nutrients and modify other properties, such as the soil pH and CEC. Some examples have been documented.

In the Dabou savannah of Côte d'Ivoire, after a first cropping cycle, oil palm plantings displayed lower yields than in the previous cycle at the same age. This situation was not caused by different cultural practices, especially fertilization; studies showed that the water supply for the replanted palms had decreased compared to the first cycle due to destructuring and compacting of the surface horizon (Caliman *et al.*, 1987). This change, which was caused by the repeated use of heavy machinery during land preparation was amplified by repeated potassium applications. Thus, enrichment of the surface horizon with exchangeable K contributed to the solubilization and precipitation of iron hydroxides and aluminium, then to the cementation of sand particles.

A study was undertaken in Ecuador based on two factorial trials testing N and K application at different rates (Dubos *et al.*, 2017b). An analysis of the soil to which the fertilizers had been applied over 10 years revealed unfavourable development of the soils when the rates exceeded the storage capacity of the soil, or N and K uptake by the crop. Nitrogen applications (urea) led to soil acidification, explained by leaching of the excess nitrogen in nitrate form. When migrating downwards, NO_3^- ions are combined with Ca^{++} and Mg^{++} cations, leading to impoverishment of the surface horizon for these bivalent cations (figure 36). In the same way, high KCl application rates caused an increase in the soil's exchangeable K (figure 37) and displacement of the Ca and Mg cations by replacement on the CEC fixation sites.

These examples show that enriching the topsoil with nutrients can have negative consequences by modifying soil properties and promoting mineral nutrient losses. To reduce such risks it is essential to avoid excessive fertilizer recommendations compared to the needs of the crop and to the ability of the soil to retain the nutrients.

Assessing soil reserves

Another worry concerns long-term changes in the mineral reserves of soils. Do those mineral reserves decrease over time if fertilizer applications are not enough to compensate for what the crop has removed?

The international organization RSPO (Roundtable on Sustainable Palm Oil)[1], as part of its certification criteria for sustainable palm oil production, recommends periodic monitoring of soil properties using appropriate analyses, every five years for example.

---

[1] RSPO: Roundtable on Sustainable Palm Oil - Table ronde sur l'huile de palme durable, https://rspo.org/

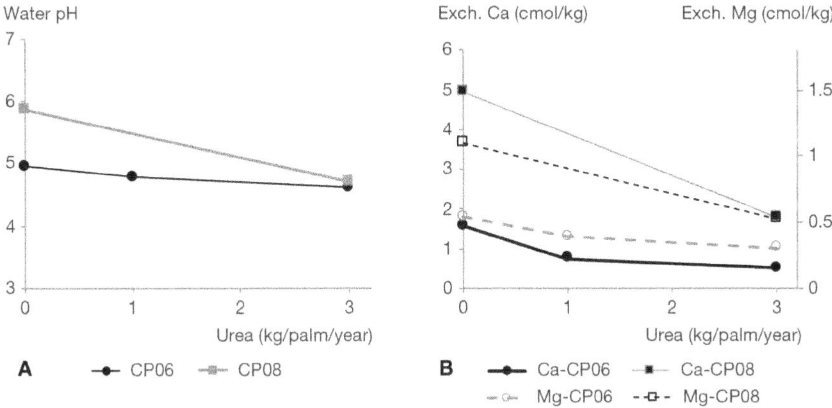

Figure 36. Measurements of topsoil pH (figure 36A) and of soil exchangeable Ca and Mg contents (figure 36B) in two factorial trials (called CP06 and CP08) implemented over 10 years in Ecuador

Soil pH, along with Ca and Mg contents, dropped when urea applications increased. The H$^+$ and Al$_3^+$ concentrations, which were also measured, significantly increased with urea. Excessive urea applications led to leaching in the form of nitrates, displacing the bivalent cations Ca^{++} and Mg^{++}.

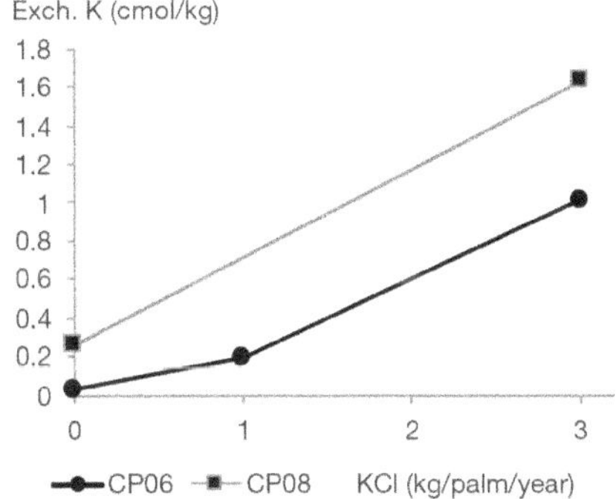

Figure 37. Measurements of exchangeable K in the soil depending on KCl applications in two factorial trials (called CP06 and CP08) implemented over 10 years in Ecuador

Potassium not taken up by the palms explains the particularly high exchangeable K contents of the soil with the application of 3 kg KCl/palm/year. That excess potassium in the topsoil is likely to migrate downwards, beyond the absorbent root layer; it can also displace the Ca and Mg cations of the CEC.

The aim is to maintain soil reserves at a sufficient level to guarantee that high yields are obtained, and soil fertility is preserved at the end of the crop cycle.

With these analyses it is possible to check retrospectively the impact of fertilization guided by leaf analysis. Nonetheless, the structure of oil palm plantings has to be taken into account, because their exploitation leads to the individualization of different soil "compartments", whose properties evolve along with the practices applied there.

For instance, these compartments are the weeded circles at the foot of the palms, harvesting tracks, zones where pruned fronds are deposited, or the areas receiving fertilizers. There are two approaches for taking this structure into account:
– assess the overall reserves in the plot by proportionally integrating each compartment. Nelson *et al.* (2015) proposed a simple method for taking composite soil samples and analysing them periodically. This method seems well adapted to long time spans (10 years, or even a whole cropping cycle),
– focus on the compartments most sensitive to impoverishment, such as unfertilized compartments or, on the contrary, the compartments most sensitive to enrichment, such as the fertilizer application areas.

Dubos *et al.* (2017b) compared soil K contents outside the fertilizer application zone in experimental plots with or without KCl application (figure 37). The aim was to measure whether the fact of not compensating for K removal by applying KCl significantly modified soil reserves compared to treatments receiving fertilizer. No significant differences were found in either of these two trials after 10 years of the protocol, suggesting that K impoverishment of the soil in the experimental plots without fertilizer application remained below the precision of the soil analyses. For factorial trial CP06, whose soil mineral reserves were low from the outset of the experiment, a K deficiency appeared with a significant drop in leaf K contents and yields (figure 29); be that as it may, the exchangeable K content of the soil in that trial did not drop to the point of being picked up by the analysis.

This example shows that by adjusting fertilizer applications based on leaf analysis results it is possible to act before soil impoverishment becomes detectable by direct analysis. This gives leaf analysis a protective role against negative effects on the soil, called "mining effects" by several authors.

Improving fertilization efficiency

Reducing mineral nutrient losses through different cultural practices

The nutrients provided by fertilizers remain at the soil surface before gradually descending into the superficial layer, where they remain until taken up by the oil palms or other plant species. During that time, several processes contribute to losses in nutrients that will not be of benefit to oil palms.

Nitrogen volatilization

Nitrogen losses through volatilization are specific to certain fertilizers, such as urea and ammonium nitrates. They generate gases (ammoniac NH_3 and nitrous oxide N_2O), which are greenhouse gases that contribute to global warming. For these fertilizers, dry periods must be avoided to reduce the risk of NH_3 volatilization. The fertilizer should be spread in the rainy season, while avoiding the wettest months as they favour the risk of N_2O volatilization, along with NO_3^- and NH_4^+ leaching.

Erosion and runoff

Nutrients are also lost through erosion, surface runoff and leaching (downward displacement of cations and nitrates beyond the soil layer explored by roots). Published data vary not only depending on the measurement sites (Dubos *et al.*, 2019), but also inside plots, as many factors are involved (climate, slope, soil cover, cultural practices). Rather than compensating for losses through fertilization that is difficult to evaluate, or overestimated as a precaution, it is wiser to try and reduce those factors as much as possible.

Erosion and surface runoff mostly depend on the slope and soil cover. These phenomena can be managed with appropriate cultural techniques (laying out pruned fronds to cover the soil, herbaceous regrowth management, cover crops) and erosion control measures (individual or continuous terraces, sediment traps).

Mineral nutrient losses through leaching

This mainly concerns N and K, due to the quantities consumed by the crop. Losses increase in sandy soils due to the low CEC. They also increase with heavy rainfall when plant cover transpiration is low and excess water is drained downwards (Banabas *et al.*, 2008). Some appropriate cultural practices reduce leaching by improving the chemical properties of the soil, which help to retain nutrients in an exchangeable form in the superficial layer:
– application of organic matter to increase the CEC. This involves pruned fronds, organic fertilizers (empty fruit bunches (EFB), fibres, compost, or other sources if available), which can be reserved for sectors where the soil has a coarse texture or where the slopes are steepest,
– spreading of mineral fertilizers over wide areas, especially where the soil has a high CEC, i.e., where organic matter accumulates,
– use of controlled-release fertilizers that release mineral nutrients without losses over 6 to 12 months and help to significantly reduce input quantities.

Lastly, splitting fertilizer rates is often recommended, even though no study has ever confirmed the effectiveness of this practice. Given the difficulty in keeping track of nutrient transfers between the soil and roots (especially K), it is not known what share of the nutrients provided by fertilizers is actually taken up between two applications and the residence time of nutrients in the soil.

Improving the physico-chemical properties of soils

Applying organic matter

Recycling available biomass is a powerful way of restoring, conserving or improving the physico-chemical properties of soils and thereby reducing mineral nutrient losses. On a plantation scale, organic fertilization improves soil fertility by inducing an increase in pH (tropical soils planted to oil palm are often acid), in the organic carbon rate, the CEC, and exchangeable cations (Comte *et al.*, 2013).

Applying organic matter brings into play complex mechanisms in the soil that go beyond a mere provision of nutrients. For example, applications of empty fruit bunches, EFB (harvesting residues from palm oil extraction) modify the communities of living organisms in the soil. Such is the case for earthworm populations, which play a crucial role in soil structure by breaking up and redistributing organic matter deep down, and by increasing aeration and water infiltration.

Organic applications promote the development of fine and absorbent roots in oil palms. When combined with a positive effect on the soil CEC this type of treatment creates conditions propitious to efficient fertilization. Over the long term, organic matter applications durably increase the production potential of vegetative biomass and bunches by regulating the plant's mineral nutrition and water supplies. This effect can be important when the climate includes a marked dry season.

Fostering atmospheric nitrogen fixation

An effective way of generating recyclable organic matter is to grow legume-based cover crops (*Pueraria* sp., *Centrosema* sp., *Mucuna* sp., etc.). This is a common practice in young crops, as it protects the soil from erosion and makes for easier weed control, especially grasses. Legumes also contribute to the nitrogen balance through atmospheric nitrogen fixing bacteria (notably of the genus *Rhizobium*), which form nodules on their roots. It is sometimes very difficult to maintain this cover once the canopy of the oil palm plantation closes, reducing light penetration. At the adult age, from 12 years onwards, the residual density of *Pueraria* sp. varies substantially from one plantation to another, and those differences in behaviour are worth looking into. It is probably not down to a simple matter of light interception by the canopy, as the regions where it is possible to maintain a plant cover sometimes coincide with low sunlight zones (e.g., western Ecuador, West Africa). When it is difficult to maintain a *Pueraria* sp. cover, shade-tolerant species such as *Desmodium ovalifolium* can be sown.

Increasing recyclable biomass

Allowing the regrowth of small woody species in adult oil palm crops is also a worthwhile way of increasing recyclable biomass. The upkeep of areas not used for harvesting and crop protection is therefore limited to the strictest minimum. Bushy dicots can be left to grow up to 2 to 3 metres in height and that biomass is pruned and returned to the soil periodically. The potential benefits of this practice are to enrich floristic diversity to promote the biological control of leaf-eating insects, increase dry matter production and recycle mineral nutrients from deeper horizons to those explored by oil palm roots. When a dry season suggests the risk of competition for water reserves, the risk can be reduced by slashing the regrowth before the problem occurs.

Developing a precise and environment-friendly fertilization tool

Each year, the challenge when fertilizing a plantation is to recommend the nutrient application rates needed to meet the requirements of the oil palms in each LSU without limiting yields. The rates recommended must also be adjusted as precisely as possible for economic and environmental reasons (carbon balance of practices, risks associated with excess N, K, etc.). There is therefore a dual need to increase the precision of the analysis and the recommendations: firstly, by defining reference contents to be achieved for each soil and each planting material and, secondly, by making recommendations that fulfil those objectives over all the areas planted.

Precision of recommendations and fertilizer schedules

In general, the precision of the recommendations and the fertilizer schedules largely depends on the quality of the soil and plant tissue analysis results. It is therefore necessary to rely on an excellent laboratory that guarantees the stability and repeatability of the results over the long term. That laboratory must be registered for soil and plant analyses and take part regularly in national or international round robin tests. It needs, in total transparency, to be able to justify its efficient operation by sharing statistical values on analysis precision (mean values and standard deviations obtained for the reference samples).

This precaution is essential for providing reliable data and constructing a decision-support tool based on leaf and soil analyses. A reference soil content is a simple and cheap synthetic indicator that guarantees a satisfactory physiological status that does not limit yield potential. That indicator is calibrated by strict and precise trials for each nutrient, depending on the climate and planting material specific to each plantation.

Leaf analysis operates over periods of around three years in a row to correct contents that deviate from the recommended values. However, agronomy trial networks sometimes show that there are interannual variations in contents that cannot be explained by the treatments applied.

They are likely due to temporal variations in the nutrient flows taken up, or to rebalancing between organs. These variations would seem to be determined by the physiological functioning of the palm and the factors that affect biomass production. It may involve a discontinuous variation in leaf biomass (deploying of several consecutive fronds for climates with a dry season), in root biomass, or in bunch production cycles. Interannual yield variations are known and probably induce nutrient flows that in turn affect leaf contents. Sudden variations that occur certain years therefore need to be interpreted with a great deal of caution.

In the future, leaf analysis will have to evolve towards a tool that takes into account these interannual variations. It will have to incorporate the variables used for yield prediction and nutrient uptake models. These are primarily the variables that act upon photosynthetic activity (climate data and soil water reserves). The shift to a generic tool with several variables will have to be backed up by robust databases (meteorological data, monthly productivity per LSU) to test hypotheses that provide access to an understanding of mineral nutrition as a yield factor.

Spatial precision of leaf sampling and fertilizer application

The first level of spatial precision for leaf analysis is leaf sampling. This is why the constitution of LSUs and the selection of palms making up the leaf samples are two key stages in obtaining a precise recommendation for a significant part of each LSU. Some special leaf samples complete this procedure to detect any drift on the minority facies of each unit. Precision can therefore be envisaged on a scale smaller than the LSU, but that means perfectly mastering specific fertilizer applications inside plots.

The spatial precision limit is that of the planted palm. Differential individual fertilizer applications depending on the geographical position of each palm is technically possible when application can be mechanized and linked up to a GPS positioning record. Such mechanization is difficult and sometimes not recommended (protection of soils) when the relief is rugged: yet, it is probably with these marked heterogeneities within plots that it seems advisable to adapt recommendations (variations in yield potentials and therefore in demand, variations in fertilization efficiency). Nonetheless, there is hope that it will become possible to delimit populations of palms inside LSUs that will receive differential fertilization if a decision-support tool becomes available that is sufficiently precise to integrate that variability.

For the time being it is not possible to access a precise recommendation for each palm. Some trials have tried analysing satellite images to convert the assumed composition of the foliage via its spectral signature into nutrient requirements, but no effective algorithm has been found to decipher the spectral values of the pixels that represent each palm. To that is added the difficulty of regularly procuring good quality images due to the abundance of cloud cover in oil palm growing regions. So, at this stage, the technology does not seem to be adapted to requirements for the time being.

Conclusion: generic tools for optimized fertilization in each plantation

The method we propose to ensure that fertilizer recommendations correspond as precisely as possible to crop requirements is a set of procedures. It involves making choices when setting up leaf sampling units (LSU) and selecting leaf and soil sampling points to monitor each unit. At the same time, experimental results have to be acquired to define the optimum content range specific to each situation. The approach is progressive, as it involves acquiring information continually on the functioning of the plantation, and making adjustments for certain areas.

However, this process cannot be applied everywhere and cannot be adopted by everybody. As things stand, it seems to be reserved for agro-industrial plantations of a given size. From around a hundred hectares or more it can be expected that annual leaf analyses will be available and that uptake tests can be implemented. From several thousand hectares, investment in one or more fertilization trials guided by an agronomist with a sound scientific grounding can rapidly prove to be cost-effective. What can be done for smaller farmers who cannot procure so much precise data for their farms?

There exist some serious prospects for developing generic tools that would take into account the specificities of each site (soil properties, climate and planting material). These tools will have to be better correlated with yields. The aim is to explain variations in leaf contents observed from one year to another, but also spatial variations (between plots, or between soil units) when they are due to biomass production cycles. These tools will have to take into account more variables than leaf analyses, especially those that govern demand linked to biomass production and those affecting fertilization efficiency. Developing robust tools can only be envisaged by modelling datasets over long periods, whether it be experimental results or plantation monitoring. Their quality, and especially the precautions taken for soil and palm sampling, will be paramount for bringing out explanatory relations and for being able to define the most efficient sets of indicators to power the tools.

There is no doubt that this work will be of help in exceeding the precision aims initially set for recommendations; it will provide an understanding of what drives productivity, as well as lessons regarding the cultural practices that promote the highest yields, especially those linked to ecosystem services offered by the soil. Such knowledge is of interest to all categories of producers, independent of their scale, and therefore provides tremendous leverage for increasing smallholders' yields, which are often lower than the potential yields achieved by agro-industries. Yet, those farmers occupy around

40% of the areas cultivated worldwide and offering them a prospect of sustainable development responds to the requirements of our era. It is about achieving our goal: satisfying global vegetable oils and fats requirements, while limiting deforestation risks and undesirable effects on water resources and local populations.

Banabas M., Turner M. A., Scotter D. R., Nelson P. N., 2008. Losses of nitrogen fertiliser under oil palm in Papua New Guinea: 1. Water balance, and nitrogen in soil solution and runoff. *Australian Journal of Soil Research*, 46(4): 332-339. https://doi.org/10.1071/SR07171

Caliman J. P., Olivin J., Dufour, O., 1987. Dégradation des sols ferrallitiques sableux en culture de palmiers à huile par acidification et compaction [Degradation of sandy ferralitic soils in oil palm cultivation through acidification and compaction]. *Oléagineux*, 42(11) : 394-401. http://agritrop.cirad.fr/453135/

Caliman J. P., Daniel C., Tailliez B., 1994. La nutrition minérale du Palmier à Huile [Oil palm mineral nutrition]. *Plantations, Recherche, Développement*, 3 : 36-54. http://agritrop.cirad.fr/387369/

Cirad-IRHO, 1969. Les symptômes de carence en magnésium chez le palmier à huile [Magnesium deficiency symptoms in oil palm. Pratique – Conseils de l'IRHO n° 80. *Oléagineux*, 24(1) : 11-12.

Cirad-IRHO, 1991. La carence azotée chez le palmier à huile. Symptômes et correction [Nitrogen deficiency in oil palm. Symptoms and correction]. Pratique – Conseils de l'IRHO n° 320. *Oléagineux*, 46(6) : 247-250.

Cirad-IRHO, 1992. La carence potassique chez le palmier à huile. Symptômes et correction [Potassium deficiency in oil palm. Symptoms and correction]. Pratique – Conseils de l'IRHO n° 332. *Oléagineux*, 47(10) : 587-591.

Cirad-IRHO, 1992. La déficience en bore chez le palmier à huile : symptômes et corrections [Boron deficiency in oil palm. Symptoms and corrections]. Pratique – Conseils de l'IRHO n° 334. *Oléagineux*. 47(12) : 719-725.

Cirad-IRHO, 1977. Préparation et conditionnement des échantillons pour le diagnostic foliaire du palmier à huile et du cocotier [Preparation and Conditioning of Samples for Leaf Analysis of Oil Palm and Coconut]. Pratique – Conseils de l'IRHO n° 170. *Oléagineux*, 32(3) : 95-99.

Cirad-IRHO, 1977. Le diagnostic foliaire pour le contrôle de la nutrition des plantations de palmier. Prélèvement des échantillons foliaires [Leaf Analysis for the Control of Nutrition in Oil Palm Plantations. Taking of leaf samples]. Pratique – Conseils de l'IRHO n° 172. *Oléagineux*, 32(5) : 211-216.

Dassou O., Nodichao L., Ollivier J., Impens R., Dubos B., Bonneau X., De Raïssac M., Durand-Gasselin T., Sinsin B., Van Damme P., 2018. Oil Palm (*Elaeis guineensis* Jacq.) Leaf K and Mg Contents Differ with Progenies: Implications and Research Needs. *In* Tielkes E. (ed.), *Tropentag 2018, Global food security and food safety: the role of Universities*, Gand, Belgique, 17-19 September 2018. Weikersheim, Margraf Publishers GmbH, 230 p. http://agritrop.cirad.fr/590883/

Dubos B., Caliman J.-P., Corrado F., Quencez P., Siswo S., Tailliez B., 1999. Rôle de la nutrition en magnésium chez le palmier à huile [Importance of magnesium nutrition in oil palm]. *Plantations, Recherche, Développement*, 6(5) : 313-325. http://agritrop.cirad.fr/264349/

Dubos B., Alarcón W. H., López J. E., Ollivier J., 2011. Potassium uptake and storage in oil palm organs: the role of chlorine and the influence of soil characteristics in the Magdalena valley, Colombia. *Nutrient Cycling in Agroecosystems*, 89(2): 219-227. https://doi.org/10.1007/s10705-010-9389-x

Dubos B., Baron V., Bonneau X., Dassou O., Flori A., Impens R., Ollivier J., Pardon L., 2019. Precision agriculture in oil palm plantations: diagnostic tools for sustainable N and K nutrient supply. *OCL*, 26(5). https://doi.org/10.1051/ocl/2019001

Dubos, B., Baron, V., Bonneau, X., Flori, A., Ollivier, J., 2017a. High soil calcium saturation limits use of leaf potassium diagnosis when KCl is applied in oil palm plantations. *Experimental Agriculture*, 54(5) : 1-11. https://doi.org/10.1017/S0014479717000473

Dubos B., Snoeck D., Flori A., 2017b. Excessive use of fertilizer can increase leaching processes and modify soil reserves in two Ecuadorian oil palm plantations. *Experimental Agriculture*, 53(02): 255-268. https://doi.org/10.1017/S0014479716000363

Dubos B., Flori A., 2014. Persistence of Mineral Fertility Carried over from the First Crop Cycle in Two Oil Palm Plantations in South America. *Oil Palm Bulletin*, 68: 8-15. http://palmoilis.mpob.gov.my/index.php/opb

Fairhurst T., Caliman J.-P., 2001. *Symptômes de Déficiences Minérales et Anomalies chez le Palmier à Huile (*Elaeis guineensis *Jacq.) : Description, Origine, Prévention, Correction* [*Oil Palm Nutrient Disorders and Nutrient Management. Diagnosis, Causes, Prevention, Treatment*]. Singapour, PPI-CIRAD-CP, 67 p. http://agritrop.cirad.fr/510748/

Foster H. L., 2003. Assessment of Oil Palm Fertilizer Requirements. *In* Fairhurst T. and Hardter R. (eds.), *Oil Palm: Management for Large and Sustainable Yields*. Potash & Phosphate Institute (PPI), Potash & Phosphate Institute of Canada (PPIC), and International Potash Institute (IPI, Basel). pp. 191-230.

Foster H. L., Prabowo N. E., 1996. *Variation in potassium fertiliser requirements in oil palm in North Sumatra*. Presented at the 1996 PORIM International Palm Oil Congress: Competitiveness for the 21st Century, Kuala Lumpur Malaysia: PORIM.

Gerendás J., Donough C., Oberthur T., 2009. Sulphur Nutrition of Oil Palm in Indonesia - The Neglected Macronutrient. *Proceedings of Agriculture, Biotechnology & Sustainability Conference*, 2: 27-31.

Goh K. J., Hardter R., Fairhurst T., 2003. Fertilizing for maximum return In Fairhurst T. and Hardter R. (eds.), Oil Palm: Management for Large and Sustainable Yields. Potash & Phosphate Institute (PPI), Potash & Phosphate Institute of Canada (PPIC), and International Potash Institute (IPI, Basel). pp. 279-306.

Jacquemard J.C., Ollivier J., Surya E., Suryana E., Permadi P., 2009. Genetic signature in mineral nutrition in oil palm (*Elaeis guineensis* Jacq.): a new panorama for high yielding materials at low fertiliser cost. *In PIPOC 2009 International Palm Oil Congress*, vol. 2, Malaysia. pp. 337-373.

Nelson P. N., Banabas M., Goodrick I., Webb M. J., Huth N. I., O'Grady D., 2015. Soil sampling in oil palm plantations: a practical design that accounts for lateral variability at the tree scale. *Plant and Soil*, 394(1-2): 421-429. https://doi.org/10.1007/s11104-015-2490-9

Ng S. K., Thamboo S., 1967. Nutrient contents in Oil Palm in Malaya 1. Nutrients required for reproduction: fruit bunches and male inflorescences. *The Malaysian Agricultural Journal*, 46(1) : 3-45.

Ollagnier M., Ochs R., 1971. Chlorine, a new essential element in oil palm nutrition. *Oléagineux*, 26(1) : 1-15.

Ollagnier M., Ochs R., 1972. Sulphur deficiencies in the oil palm and coconut. *Oléagineux*, 27(4): 193-198. Retrieved from Oléa.

Ollagnier M., Ochs R., 1981. Management of mineral nutrition on Industrial Oil Palm Plantation - Fertilizer Savings. *Oléagineux*, 36(8-9): 409-421.

Rincón Numpaque A. H., Torres Aguas J. S., 2015. Extracción de nutrimentos en racimos de palma híbrida *Elaeis oleifera × Elaeis guineensis* (O × G) en la Zona Suroccidental. Estudio preliminar [Nutrient extraction in bunches of hybrid palm *Elaeis oleifera × Elaeis guineensis* (O × G) in the Southwestern Zone. Preliminary study]. *Ceniavances*, (182) : 1-8.

Tampubolon F. C., Daniel C., Ochs R., 1989. Oil palm responses to nitrogen and phosphate fertilizer in Sumatra. Presented at the 1989 PORIM International Palm Oil Development Conference, Kuala Lumpur. pp. 419-428.

Tan G. Y., Rajaratnam J. A., 1978. Genetic variability of leaf nutrient concentration in oil palm. *Crop Science*, 18 (July August): 548-550.

Tarmizi A. M., Mohd Tayeb D., 2006. Nutrient Demands of Tenera Oil Palm Planted on Inland Soils of Malaysia. *Journal of Oil Palm Research*, 18(1): 204-209.

Teoh, K. C., Chew P. S., 1987. Potassium in the oil palm ecosystem and some implications to manuring practice). Presented at the 1987 International Oil Palm / Palm Oil Conferences: Progress and Prospects, Kuala Lumpur, Malaysia. pp. 277-286.

Webb, M. J., 2009. A conceptual framework for determining economically optimal fertiliser use in oil palm plantations with factorial fertiliser trials. *Nutrient Cycling in Agroecosystems*, 83(2): 163-178. https://doi.org/10.1007/s10705-008-9207-x

Webb M. J., Berthelsen S., Nelson P., van Rees H., 2009. Final Report. Overcoming magnesium deficiency in oil palm crops on volcanic ash soils of Papua New Guinea. Canberra, ACIAR, Report FR2009-11. https://www.aciar.gov.au/node/11486

Yates M., 1964. The Design and analysis of factorial experiment. Commonwealth Agricultural Bureaux, Bucks, 98 p. Public Library of India https://archive.org/details/in.ernet.dli.2015.449111/page/n1

Layout: Hélène Bonnet – Studio 9

Imprimé pour vous par Books on Demand (Allemagne)